Your AI Isn't Broken.
Your Playbook Is.

For leaders who greenlight AI. For practitioners who make it work.

AI project failures don't come from bad models. They come from old habits. Lock AI down like software and you kill its value. Let it roam and it drifts. Mistakes ship and scale. WISER is the field manual for teams operationalizing AI.

We asked four frontier AI models to review this book. Their assessments are below. Your job, the one AI can't do, is to discern how and when to trust their output. That collaboration is what this book is about.

> "The Lean Startup of the AI era." - Gemini
> "Belongs on the shelf alongside The Phoenix Project." - ChatGPT
> "A field manual from people who've shipped systems." - Claude
> "Sobriety, structure, and hard-won wisdom." - xAI

What You Walk Away With:
The complete WISER Method. Built on AI First Principles (aifirstprinciples.com). Taught through cognitive layering that builds your mental model as you read. Close the book and you are ready to move.

- **Real-World Narrative** following one company from discovery to deployment
- **26 Plays** with objectives, inputs, steps, success criteria, adaptations, and pitfalls
- **20+ Downloadable Templates** ready to use on day one
- **Advanced Frameworks** for risk, autonomy, measurement, and knowing when to rebuild

Robb Wilson & **Anthony Franco** have been building together for twenty years. They started with Effective, one of the first design thinking firms, and sold it to WPP. Robb went on to found OneReach, an enterprise AI platform, and write the bestselling *'Age of Invisible Machines'*. Anthony founded and sold five tech startups. Between them, they've built systems for over half the Fortune 100.

They built this playbook for themselves. **Now it's yours.**

wisermethod.com/master_playbook

Author's Note

You're about to learn a method for building AI systems. But the method won't stick if we just hand it to you. You need the thinking that makes it work.

So we structured this book using cognitive layering. Each chapter introduces one concept that the next chapter builds on. The progression is deliberate. By the time you reach the WISER Method, you'll have the mental models to actually use it. By the time you reach the case study, you'll recognize the patterns. By the time you finish, you won't just know what to do. You'll know why it works.

We wrote the book using the same approach we use in the field. We observed where practitioners got stuck, patterns we recognized from actual implementations. We tested which concepts needed more foundation. We iterated until the progression felt inevitable. The WISER Method shaped the book about the WISER Method.

We wrote this book as the workbook for the WISER Method course, but it stands on its own. The course adds facilitated practice and guided implementation. Everything you need to understand and apply WISER is here.

Who This Book Is For

Leaders evaluating how to implement AI: CAIOs, CTOs, COOs, and CEOs who need to understand what good implementation looks like so they can sponsor it, fund it, and recognize when it's going wrong. Your path is Parts I, II, and III; roughly 80 pages. That gives you enough to sponsor well and recognize when things go wrong. Part IV is the practitioner toolkit.

Practitioners who make change stick: consultants, internal champions, process improvement leads, scrum masters, and anyone tired of watching promising pilots die in production. The entire book is your workbook. Part IV is where you'll spend most of your time in the field.

One caution before you begin: WISER requires conditions to succeed: executive buy-in, psychological safety, room to experiment. Not every organization can provide them. If you're uncertain whether the method fits your situation, Chapter 20 names where WISER fails.

Robb & Anthony

WISER Method

Master Playbook

Build What Matters with AI First Principles

Robb Wilson

Anthony Franco

First Strategy in association with UX MAG

firststrategy.ai | uxmag.com

Copyright 2026 by Anthony Franco and Robb Wilson

All rights reserved.

No part of this publication may be reproduced, distributed, or transmitted in any form or by any means, including photocopying, recording, or other electronic or mechanical methods, without the prior written permission of First Strategy, LLC, except in the case of brief quotations embodied in critical reviews and certain other noncommercial uses permitted by copyright law.

ISBN: 979-8-218-93145-2

LCCN: 2026902988

First Edition

Published by First Strategy in association with UX MAG

firststrategy.ai | info@firststrategy.ai

Book Design and Illustrations: Anthony Franco

Printed in the United States of America

10 9 8 7 6 5 4 3 2 1

Acknowledgments: Deep gratitude to Catherine Joss, Elias Parker, Josh Tyson, Sasha Wilson, and Shawna Franco, whose sharp editing, critical feedback, and unwavering support shaped this book from rough drafts to finished pages.

Disclaimer: The information in this book is provided for educational and informational purposes only. The authors and publisher make no representations or warranties with respect to the accuracy or completeness of the contents. The case study company "Wingman Services" is a composite illustration. AI First Principles is an open-source framework at aifirstprinciples.com.

About the Reviews: The reviews on the back cover were generated by prompting each frontier AI model as an experienced business book critic. Full reviews available at wisermethod.com/playbook.

WISER Method is a trademark of First Strategy LLC.

Contents

Part I: The Problem 9
    Chapter 1: The Jungle 11

Part II: The Worldview 15
    Chapter 2: Why AI Is Different 17
    Chapter 3: The WISER Perspective 23
    Chapter 4: The WISER Method 27

Part III: The Case Study 37
    Chapter 5: Witness 39
    Chapter 6: Interrogate 47
    Chapter 7: Solve 55
    Chapter 8: Expand 63
    Chapter 9: Refine 75

Part IV: Starter Plays 89
    Chapter 10: Team Setup 91
    Chapter 11: Playbook System 101
    Chapter 12: Witness Plays 109
    Chapter 13: Interrogate Plays 121
    Chapter 14: Solve Plays 137
    Chapter 15: Expand Plays 147
    Chapter 16: Refine Plays 163
    Chapter 17: Rhythm Plays 185

Part V: Mastery 197
    Chapter 18: Advanced Frameworks 199
    Chapter 19: Creating Your Own Plays 207
    Chapter 20: When WISER Fails 213
    Chapter 21: Integrating Methods 219
    Chapter 22: The Journey Continues 233

Appendices
    Appendix A: Parking Lot 239
    Appendix B: References 245
    Appendix C: Index 249
    Appendix D: Glossary 253

PART I

Part I: The Problem

I spent decades mastering how technology projects work. Requirements, timelines, discipline, rigor. I was good at it.

I wasn't wrong. The world just stopped being the place where I was right.

- *Anthony*

PART I

Chapter 1: The Jungle

In May 2023, a small plane crashed in the Colombian Amazon.

On board were four children and three adults. All three adults died, including the children's mother. The oldest child was thirteen. The youngest was one.

They were alone. Lost. In one of the most dangerous environments on Earth.

For forty days, they survived. Waiting.

The Colombian military launched the largest search operation in the country's history. Hundreds of soldiers alongside dozens of indigenous trackers, including Huitoto people who knew the jungle differently. Helicopters. Heat-sensing drones. Grid-search protocols refined over decades of jungle warfare, combined with generations of jungle knowledge. They covered hundreds of square miles, methodically, systematically.

For thirty-five days, they found nothing.

Then an elder on the search team did something no protocol would sanction.

He drank yagé.

Not for recreation. For vision. A plant medicine his people had used for guidance for centuries. That night, he told the search team: we will find the children today.

The next morning, they walked into a clearing. The children were there.

Where would a thirteen-year-old hide when he hears a helicopter but doesn't know if it's rescue or danger? Thirty-five days of grid searches and jungle tracking couldn't answer that question. One vision could. It reached past protocol into something precision had no room for: the jungle as the children experienced it, not as the operation mapped it.

Thirty-five days of the best search in Colombia's history. One night that changed where to look.

That story haunts me.

I've spent thirty years building technology companies. Seven startups. Six acquisitions, two to public companies. More Fortune 100 consulting engagements than I can count. I know how projects work. Requirements, timelines, resources, milestones. The discipline of execution. The comfort of a plan.

That's how I was conditioned to operate. Precise. Disciplined. Governed. Proud of the rigor.

The failures I've witnessed all followed the same pattern. Smart teams. Real budgets. Genuine commitment. They did everything right. They gathered requirements. Built specifications. Followed the waterfall or sprinted through the agile motions. They executed the search plan with precision.

They found nothing.

Or worse: they found something that looked like success but delivered no value. A chatbot nobody uses. A recommendation engine that recommends what customers already bought. A prediction model that's accurate in testing and useless in production. Pilots that never graduate. Proofs of concept that prove nothing except the team's ability to build demos.

The helicopters flew. The grid got covered. The dashboards glowed green.

And the value stayed hidden.

This book is about finding that value. Not the metrics. Not the KPIs. Not the executive dashboards that prove activity happened. The actual value AI can deliver. The transformed workflows. The delighted users. The genuine expansion of what your organization can accomplish.

The jungle is real. AI implementation is genuinely complex. Requirements shift because the technology learns and changes. Stakeholders don't agree on what success looks like. Edge cases multiply. The thing you built yesterday behaves differently today. Uncertainty isn't a bug in the process; it's the defining characteristic of the territory.

Most organizations respond to this complexity by doubling down. More requirements. Longer specifications. Tighter controls. Bigger governance committees. They're flying more helicopters over the same ground, measuring progress in activity instead of outcomes.... It doesn't work.

The search didn't fail for lack of method. It failed for lack of the right method. Thirty-five days of grids and protocols searched where trained adults would go. The elder's vision searched where frightened children would hide.

To find a scared child in a jungle, training isn't enough. You need empathy. Experience isn't enough. You need imagination. A plan for every contingency isn't enough. You need the ability to observe, adapt, and respond to what's actually happening on the ground.

This is what WISER teaches.

Not the absence of discipline. A different discipline. One designed for the actual territory you're operating in. Probabilistic systems that learn and change. Stakeholders whose needs evolve as they see what's possible. Outcomes that can't be specified in advance because the real value only becomes visible through iteration.

The military wasn't wrong to search. They were wrong about how to search. Their methodology was built for a different kind of problem. Fixed targets. Known locations. Predictable movement. The children weren't a fixed target. They were moving, hiding, surviving in ways that adult training couldn't predict.

Your AI implementation isn't a fixed target either.

I wrote this book with Robb Wilson.

Robb wrote *Age of Invisible Machines* and has spent two decades building AI systems for Fortune 100 companies and federal institutions. He's one of the founding contributors to AI First Principles. Where I bring the operator's perspective, Robb brings the builder's eye for what actually ships.

Between us, we've implemented systems in organizations ranging from Fortune 100 enterprises to federal institutions. We've seen what works. More instructively, we've seen what fails, and we've noticed the pattern.

The failures look like the grid search. Disciplined. Thorough. Proud of the rigor. And consistently missing the point.

The successes look like the collaboration. Traditional capability combined with a different way of thinking. Observing before acting. Testing assumptions instead of defending them. Iterating toward value instead of executing toward a specification. Building empathy before building solutions. Understanding humans before automating their work.

Robb and I built one of the world's first design thinking firms, Effective, and sold it to WPP. For two decades, I watched companies treat human-centered design as optional. Most skipped it. The ones who didn't built better products.

With AI, the math has changed. Empathy isn't optional anymore. The companies winning at AI aren't the ones with the best models. They're the ones who understand the humans those models serve. There's a reason OpenAI acquired Jony Ive's company.

We built WISER to teach teams how to combine traditional capability with this way of thinking.

Here's what's coming.

Chapter 2 explains why AI is operationally different. The rules that worked for deterministic software don't work for probabilistic systems. Chapter 3 presents the worldview that makes WISER possible: scarcity or abundance, and why that choice determines everything that follows.

Chapter 4 introduces WISER itself. The principles. The structure. The five Canons that take you from observation through refinement.

Chapters 5 through 9 follow one company through all five Canons. We call them Wingman Services, founded by a former Navy fighter pilot. The company is a composite — the patterns come from real implementations we've led. Identifying details have been changed. They thought they needed better scheduling software. What they actually needed was a complete rethinking of how their back office operated.

Chapters 10 and 11 set up your team and documentation. Chapters 12 through 17 give you the Plays. Positions to fill. Templates to use. Cadences to follow. The practical core.

The final chapters address mastery: advanced frameworks, creating your own Plays, and where to go from here.

The children in Colombia were found because someone combined traditional capability with a radically different way of thinking.

Your organization's value is out there. The helicopters have been searching. The grid has been covered.

Maybe it's time to think like the people who actually know where to look.

Part II: The Worldview

The jungle story is the feeling. What follows is the thinking that makes it work.

These three chapters build the foundation. Chapter 2 explains why AI breaks the rules that governed software for fifty years. Chapter 3 introduces the perspective that replaces those rules. Chapter 4 lays out the method itself: the Canons, the Playbook, and the Positions that make it work.

By the end of Part II, you'll have the mental model. Part III shows you what it looks like in practice.

PART II

Chapter 2: Why AI Is Different

In August 2025, Andrej Karpathy posted this on X:

> I get ~10 spam calls per day (various automated voicemails, "loan pre-approval" etc) and ~5 spam messages per day (usually phishing).
> - I have AT&T Active Armor, all of the above still slips through.
> - All of the above is always from new, unique numbers - blocking doesn't work.
> - I am on all Do Not Call lists.
> - I have iOS "Silence Unknown Callers" on, but even if it catches & silences them I still get the notifications.
>
> Not sure if other people are seeing something similar or figured out anything that works.

Two million people saw that post.

Karpathy isn't some random guy complaining about spam. He's the former Tesla AI chief. OpenAI founding researcher. The person who literally built the technology powering ChatGPT. If anyone should have AI figured out, it's him.

And he's drowning.

The technology he helped create is now deployed by anyone with an internet connection. Some weaponize it. Others struggle to harness it. Most organizations don't know what to do with it at all.

That gap—between what AI can do and what organizations can manage—creates pressure. Pressure creates panic.

Panic looks like this: a CEO reads about ChatGPT, tells the CTO to "do something with AI," and six months later there are fourteen disconnected pilots, three competing vendor contracts, and a PowerPoint deck claiming transformation while nothing has actually transformed.

We call this panic buying. Point solutions acquired without strategy. Disconnected tools that don't talk to each other. Vendor-driven adoption where the roadmap belongs to someone who doesn't work for you.

Bolting AI onto broken processes scales the mess.

If your customer service process requires three handoffs and two days to resolve a simple issue, adding an AI chatbot doesn't fix the process. It gives customers a faster way to discover how broken you are. The AI doesn't hide the dysfunction. It amplifies it.

Organizations that panic-buy end up with impressive demos and disappointing results. They confuse activity for progress. They measure success in pilots launched instead of value delivered.

The alternative isn't slower adoption. It's smarter adoption.

The question facing every organization right now is not whether to adopt AI. That question is already answered. The question is how to adopt AI in a way that builds capability instead of technical debt.

This requires understanding something uncomfortable: AI development is fundamentally different from the software development most organizations know.

The rules that worked for twenty years of enterprise software don't work here. Not because they were wrong. Because the territory changed.

What You Knew	What's True Now
Spec → Build → Test → Ship	Observe → Hypothesize → Test → Iterate
Requirements define success	Evidence defines success
Edge cases are bugs to fix	Edge cases reveal system boundaries
Testing proves it works	Testing proves how it fails
Documentation captures truth	Documentation captures assumptions
Failure is process error	Failure is expected learning
Experience is an asset	Experience can be a liability

That last row may be the most difficult to accept. The difference comes down to how you build.

Traditional software is deterministic. You tell the system exactly what to do. If the user clicks this button, show that screen. If the account balance drops below zero, reject the transaction. Every behavior is specified. Every outcome is designed. The system does precisely what you told it to do, nothing more, nothing less. Same input, same output, every time.

This is how most organizations learned to build software. Write requirements. Specify behaviors. Code the logic. Test that the code matches the spec. Ship.

AI systems are probabilistic. You don't tell the system what to do. You give it an objective and guardrails. The objective says what you're trying to achieve. The guardrails say what the system is not allowed to do. Within those constraints, behaviors emerge from the system's training and the patterns it has learned. You shape the outcomes, but you don't script them.

This is a fundamental inversion. Deterministic development specifies the positive space: here's exactly what should happen. Probabilistic development specifies the negative space: here's what must not happen, now find a path through everything else.

The mental shift is profound. You're no longer the architect of every behavior. You're the designer of constraints within which behaviors emerge.

In a widely cited 2017 essay, the same Karpathy articulated this shift better than anyone:

> "In Software 1.0, the code is written by humans. In Software 2.0, the code is written by optimization. We have shifted from being coders to being curators of goals and data."

Curators, not coders. Architects, not algorithms.

We do three things AI can't:

Aspire. Decide what's worth pursuing. AI optimizes toward goals. Humans choose purpose.

Discern. Judge quality, truth, and fit. AI generates options. Humans recognize which ones are right.

Create. Generate genuinely new ideas. AI recombines existing patterns. Humans break them.

This is the new division of labor. AI handles scale and speed. You handle aspiration, discernment, and creation.

In deterministic software, experience compounds. Someone who's built payroll systems for fifteen years has seen every edge case, every integration nightmare, every regulatory change. Their experience predicts future problems before they happen. You want that person on your team.

In probabilistic systems, experience can mislead.

The person who knows exactly how software "should" behave gets frustrated when AI systems behave differently each time. The expert who's mastered requirements gathering gets stuck when requirements can't be specified in advance. The project manager who's shipped fifty waterfall projects keeps looking for the requirements phase that doesn't exist.

This isn't about intelligence or adaptability. It's about pattern matching. When your brain has twenty years of "this is how software works" patterns, it takes effort to see that those patterns don't apply anymore.

Experience isn't worthless. But experience without the ability to unlearn is dangerous.

The operational implications stack up fast.

Same input, different output. Run the same prompt twice, get different results. Traditional testing assumes repeatability. AI doesn't offer that.

It changes while you're not looking. The system you deployed last month learned from what it encountered. Your documentation is already wrong.

Behaviors nobody designed. Useful ones. Dangerous ones. You can't predict them from the spec because they weren't in the spec.

Context breaks everything. What works in Denver fails in Phoenix. The model learned patterns from one environment and those patterns don't transfer cleanly. Scaling isn't copying.

Black box decisions. You often can't explain why the system did what it did. You can observe patterns. The mechanism stays hidden.

These aren't flaws. They're the technology.

Cassie Kozyrkov, Google's former Chief Decision Scientist, put it simply:

> "In traditional programming, the human provides the recipe (the code) and the computer follows it to make the dish. In machine learning, the human provides the dish (the examples) and the machine figures out the recipe."

You don't get to negotiate with these characteristics. You only get to decide whether you're prepared for them.

Organizations that succeed with AI share a pattern: they stopped applying deterministic methods to probabilistic problems.

They stopped writing specifications that assume they know the answer before building. They started treating every deployment as an experiment. They built systems that expect to be wrong and can recover gracefully. They measured success by outcomes, not activities.

They also stopped trying to manage AI projects like traditional projects.

The Gantt chart that worked for your ERP implementation doesn't work here. The phase gates that caught problems in waterfall become bottlenecks that slow learning. The project manager role that coordinated deterministic work doesn't match the iteration speed that probabilistic systems require.

This isn't a criticism of project management. It's recognition that tools designed for one environment don't automatically work in another.

The tsunami is real. The pressure to adopt is real. The risk of panic buying is real.

But so is the opportunity. Organizations that figure out how to build with AI, not just buy AI tools, will have advantages that compound. Each successful implementation teaches them something. Each lesson makes the next implementation faster. The gap between organizations that learn and organizations that buy grows wider every month.

For organizations with competitors who are moving, stagnation is more dangerous than change.

The teams maintaining systems that competitors are replacing feel safe. They're not. They're accumulating technical debt while the market moves. Every month of "let's wait and see" is a month their competitors are learning and their customers are arming themselves with AI that never sleeps. The gap compounds on both sides.

The question is not whether to move. The question is how to move without panic. How to adopt AI in a way that builds real capability instead of impressive demos. That requires a different way of thinking.

It starts with a choice: scarcity or abundance.

Chapter 3: The WISER Perspective

AI eliminates bureaucracy, not work.

That sentence separates organizations that thrive with AI from organizations that merely survive it.

The question isn't whether AI will change how your organization operates. It will. The question is what you do with the capacity it liberates. You have two choices:

Scarcity. Use AI to do the same work with fewer people. Optimize for headcount reduction. Measure success in costs cut. This works in the short term. Spreadsheets improve. Shareholders applaud. And then stagnation sets in, because you've used a transformative technology to do less, not more.

Abundance. Use AI to tackle bigger problems with more capacity. Redeploy the people whose routine work got automated into work that requires judgment, creativity, and human connection. Measure success in capability expanded. This is harder to explain to a board. It's also how you build something competitors can't copy.

The highest-value companies choose expansion. NVIDIA, Tesla, Epic, ASML. They don't get there by cutting costs. They organize around flat structures, meritocracy, and human autonomy. They give lateral-thinking teams hard problems and the freedom to solve them. These teams attack challenges from angles others miss, finding paths that don't exist on the standard playbook. That organizational model creates a competitive moat that efficiency-focused companies can't replicate.

AI makes that organizational model accessible to companies that couldn't achieve it before. The structure that built NVIDIA's advantage can now be replicated more readily. Abundance isn't just a philosophy. It's becoming operationally achievable.

The scarcity mindset is seductive because it's measurable.

Fire ten people, save their salaries. The ROI is obvious. The board understands it. The CFO can model it. Nobody has to explain what "expanded capability" means or how to measure it.

But scarcity has a ceiling. Once you've automated the routine work, you're done. You've captured the efficiency gains. Now you're a smaller organization doing the same things, competing against organizations that used the same technology to do bigger things.

Abundance is harder to execute because it requires imagination. If we automate this process, what could those people do instead? What problems have we been too resource-constrained to tackle? What markets have we been too slow to enter? What customer needs have we been too busy to notice?

These questions don't have spreadsheet answers. They require leadership willing to invest in capability before the ROI is proven.

The abundance path requires a different way of working, not just a different goal.

Consider two companies facing the same problem: customer service teams drowning in repetitive inquiries.

Company A spent six months on requirements. They built exactly what the spec described. On launch day, reps refused to use it. The AI answered questions customers weren't actually asking. The documented process had drifted from reality years ago; nobody had bothered to watch the reps work.

Company B spent two weeks observing. They sat with reps and noticed what nobody had documented. Then they built something small: a single feature handling the three most common questions. Reps loved it because it solved their actual pain. The team iterated weekly, adding capabilities as they proved value. Six months later, they were handling 40% of inquiries automatically. Not because they planned it. Because they earned their way there.

Same problem. Same technology. Different outcomes.

The difference wasn't intelligence or budget. It was how they thought about the work.

The perspective that makes WISER possible rests on three tenets.

Action over Theory. Trust what can be proven, not what can be planned.

Traditional project management assumes you can specify success in advance. Define requirements, estimate effort, build to spec. This assumption has always been shaky. The companies we admire most, the NVIDIAs and Teslas, rejected this model long before AI entered the conversation. They built through iteration, not specification. They trusted demonstrated results over projected outcomes.

AI makes the traditional model's weakness impossible to ignore. You don't know what the system will do until you build it. You don't know what users will do with it until they use it. Plans that assume certainty become expensive fiction faster than ever.

Action over Theory means replacing planning with proving. Small experiments that reveal what works. Evidence that accumulates into confidence. Decisions made on demonstrated reality.

Evolution over Disruption. Rebuild the system while it runs, not shutting it down for a rewrite.

The fantasy of the clean-slate rebuild is persistent and dangerous. Tear it all down. Start fresh. Build it right this time.

Organizations can't pause operations to rebuild from scratch. The business keeps running. Customers keep calling. Revenue keeps mattering. Any methodology that requires stopping the world to improve it isn't a methodology. It's a fantasy.

Evolution means advancing what's broken without stopping what works. Bounded improvements that reveal system behavior. Components that can be replaced independently. Not revolution. Evolution.

People over Proxies. Value the experts doing the work over the ones documenting it.

Documentation lies. Not intentionally, usually. But the gap between what's documented and what's actually happening grows wider every day. The official process describes how things were supposed to work three years ago. The actual process lives in the heads of the people doing the work.

People over Proxies means treating the humans doing the work as the primary source of truth. Not the project managers describing what those people do. Not the business analysts who interviewed them once. The people who know where the friction is because they feel it every day.

These tenets aren't inspirational posters. They're operational constraints.

Action over Theory means you don't get to plan for six months before building. Your first implementation will be wrong, and that's not failure, it's learning. The team that ships fast and iterates beats the team that plans perfectly and ships late.

Evolution over Disruption means you don't get to propose the clean-slate rebuild. Every change has to work alongside what's already running. Complexity is managed, not eliminated.

People over Proxies means you don't get to design from the conference room. You talk to the dispatcher, the technician, the billing clerk. Your carefully constructed process map is probably wrong. Watch what actually happens before you optimize it.

These constraints feel limiting. They are. That's the point. Constraints that prevent common failure modes are features, not bugs.

The scarcity path and the abundance path look similar at the start.

Both automate routine work. Both deploy AI to handle repetitive tasks. Both generate efficiency gains. The divergence happens after the initial deployment, when leadership decides what to do with the liberated capacity.

Scarcity organizations bank the savings and declare victory. Abundance organizations reinvest the capacity and ask what's next.

Most organizations will fund abundance through scarcity. The early efficiency gains buy the room to experiment. The choice isn't which path to walk; it's which path to walk toward.

The gap compounds. Scarcity organizations get incrementally more efficient. Abundance organizations get fundamentally more capable. Three years later, they're not even competing in the same category anymore.

This is the perspective that makes WISER possible.

Action over Theory. Evolution over Disruption. People over Proxies.

The organizations that thrive chose abundance. The next chapter shows you how they build.

Chapter 4: The WISER Method

> "It's never the genie that's dangerous. It's the unskilled wisher."
> — Cassie Kozyrkov

AI is the genie. Traditional project management is the unskilled wisher. The wisher learned to manage people, processes, and predictable systems. The genie operates on different physics. It fails silently, learns from mess, scales bias, and drifts without warning.

The wisher keeps asking for what worked before. The genie keeps delivering exactly what was asked for. The results keep disappointing.

The gap isn't effort. It's orientation. The wisher lacks the principles required to instruct a probabilistic system. Without those principles, every wish is a gamble.

AI First Principles closes that gap. It's an open-source framework of twelve constraints for building AI systems that work. Contributed by dozens of practitioners, researchers, and builders who've spent decades implementing AI systems that work and autopsying systems that don't, the principles are refined continuously by the community that uses them (aifirstprinciples.com). These aren't suggestions. They're the physics of AI implementation. Violate them and your system fails in predictable ways.

The WISER Method translates the principles into practical motion. The principles tell you what's true. The method shows you what to do about it.

What follows is both: the twelve principles that govern AI implementation, then the method that operationalizes them.

The 12 AI First Principles

AI Inherits Messiness

AI learns from people. Therefore AI systems, like people, are inconsistent and operate more effectively with structure. Trying to engineer them to operate like deterministic code will result in system failures. Variation is inevitable, not accidental.

The constraint: Define what's prohibited over what's required.

AI Fails Silently

AI accumulates errors across thousands of interactions before patterns become visible. Traditional systems failed loudly with clear signals; AI fails quietly on repeat.

The constraint: Build feedback loops over post-mortems.

People Own Objectives

AI shouldn't be used in place of human discernment, judgment, or taste. When AI makes mistakes or causes harm, a person should be held accountable, not the algorithm.

The constraint: Name the owner.

Deception Destroys Trust

When AI pretends to be human, people cannot calibrate their expectations, recognize its limitations, and protect themselves from its failures.

The constraint: Make AI obvious, not hidden.

Individuals First

AI industrializes manipulation by personalizing it at scale. Mass persuasion becomes individual manipulation. Build tools that people control, not tools that control people.

The constraint: Prioritize individual agency above efficiency, profit, or convenience.

Build from User Experience

Without input from end users, AI solutions won't solve real problems. People wrestling with system failures are the ones qualified to design system futures.

The constraint: Design systems from lived experience, not distant observation.

Discovery Before Disruption

Changing systems that aren't understood creates unpredictable failures. Redundancies prevent edge cases and manual steps catch exceptions. Existing inefficiencies are containers of knowledge.

The constraint: Identify purpose before simplifying.

Ambiguity Is Wisdom

Concealing ambiguity removes opportunities for critical judgment. Not all decisions are binary yes/no. Wisdom lives in gray areas. AI produces probabilities that demand judgment, not facts that replace it.

The constraint: Surface the probabilities.

Reveal the Invisible

There's a wealth of ignorance hiding in document theater. Expose what you don't yet understand by learning how to articulate it.

The constraint: Pursue what is hard to explain.

Iterate Towards What Works

Grand plans commit to solutions without validating problems. Iteration tests assumptions and measures impact, revealing what works gradually over time. Inherited practices carry outdated logic that meetings can't expose.

The constraint: Learn by doing, not planning.

Decompose Incrementally

Legacy systems carry too much technical debt to replace and are too brittle to automate. AI systems should allow isolated components to decompose naturally.

The constraint: Dismantle legacy complexity piece-by-piece.

Justify Resource Consumption

AI makes it trivially easy to waste resources. What costs pennies to create can cost millions to run. The friction that once prompted resource consideration has vanished. The resources remain real: energy, water, compute, time.

The constraint: Optimize the ratio of value per resource spent.

Twelve principles. Twelve constraints extracted from systems that failed and systems that held. They form the foundation. What follows is the structure for building on it.

The WISER Method

WISER enables teams to innovate continuously without operational disruption. The method produces four capabilities:

- **Continuous evolution:** Rebuild systems while they run
- **Systematic risk burn-down:** Identify and reduce risk through evidence
- **Living documentation:** Memory that survives the chaos
- **Clear ownership:** Every decision has a person accountable, even as AI scales

The Problem WISER Solves

WISER is built for teams where stagnation is more dangerous than change. Where maintaining systems that competitors are replacing feels like the safe choice. Where changing things feels riskier than living with dysfunction. Where opportunities pass because moving feels too risky.

AI can obliterate that stagnation, but only if teams dismantle bureaucracy rather than automate it. Automating dysfunction just produces faster dysfunction. Since organizations can't pause operations to rebuild from scratch, they need a way to advance that fixes what's broken without stopping what works.

How WISER Works

WISER builds perpetual innovation through systematic risk burn-down: identify the highest-risk items, reduce them through evidence and iteration, move to the next. Bounded improvements reveal system behavior, validate what works, expand capability.

Capability creates decisions. Decisions drive action. Action expands capability. The cycle feeds itself when you have structure to prevent chaos.

The Structure

WISER operates as a cohesive system, not a linear checklist.

Canons drive the strategic momentum, moving from observation to scale.

Plays adapt that strategy to the specific reality of the domain.

A **Playbook** enforces explicit constraints, serving as living documentation that prevents organizational amnesia.

Positions assign human accountability for critical decisions.

The result is a self-correcting engine that scales innovation while managing risk deliberately.

The Five Canons

The W-I-S-E-R Canons build organizational capacity to innovate continuously without rebuilding from scratch. They reflect what works when teams need to evolve systems that can't shut down.

W: Witness

Observation reveals what planning conceals.

This Canon begins here because documentation theater often hides the workarounds and hacks that keep systems running. Optimizing based on the official process means optimizing fiction. Witness demands mapping the friction people actually feel, forcing the solution to address real problems rather than theoretical ones.

The principles at work: Build from User Experience. Reveal the Invisible. Discovery Before Disruption. Deception Destroys Trust.

I: Interrogate

Observation finds pain. Experiments find causes.

This Canon exists to avoid the most common failure mode: building the wrong solution perfectly. Instead of committing to months of development, rapid experiments reveal root causes. The goal is not to guess what's broken, but to force the system to reveal it.

The principles at work: Iterate Towards What Works. Reveal the Invisible. Build from User Experience. AI Inherits Messiness. Ambiguity Is Wisdom. Deception Destroys Trust.

S: Solve

Experiments find causes. Solutions earn trust.

The focus is on delivering a single, working solution that demonstrates undeniable value. Working software settles arguments. This approach secures the organizational permission required to touch critical systems by delivering a win that matters.

The principles at work: Iterate Towards What Works. Reveal the Invisible. Build from User Experience. Justify Resource Consumption. People Own Objectives. Deception Destroys Trust.

E: Expand

Earned trust enables systematic change toward autonomy.

Modularizing the successful component allows it to solve related problems while maintaining explicit human oversight. This scales the solution's reach without introducing the systemic risk that comes from all-or-nothing deployments.

The principles at work: Decompose Incrementally. Reveal the Invisible. Build from User Experience. Justify Resource Consumption. AI Fails Silently. People Own Objectives. Deception Destroys Trust.

R: Refine

Autonomy is not designed, it is grown.

AI autonomy increases as reliability is proven. Trust is earned, not designed. Agency transfers to the system as it proves it can make decisions correctly without breaking the boundaries defined in your Playbook.

The principles at work: AI Inherits Messiness. Reveal the Invisible. Build from User Experience. Justify Resource Consumption. Decompose Incrementally. AI Fails Silently. Deception Destroys Trust.

The Canons follow a logical sequence, but iteration is expected, not failure. Discovering new information while in Expand may require returning to Interrogate to test assumptions. This scaffolded flexibility (structure that permits discovery) is what enables lateral thinking at scale without descending into chaos.

Plays

Plays transform the abstract WISER framework into tactical execution for specific contexts. They give practitioners proven patterns instead of blank pages.

A Play is not a rigid prescription. It's a starting point for adaptation. Each Play is a modular component: a mapping technique, a framework, a template, a cadence. You can use Plays as-is, rename them, combine them, or replace them with alternatives that fit your context.

This modularity is intentional. A startup might use the standard Positions Play but replace the Playbook template with something lighter. A regulated industry might keep the Playbook but add compliance-specific sections. A consultant might develop a "Healthcare AI Play" that adapts several components for that domain.

Part IV of this book provides the Starter Plays: a collection of Plays that work together but are independently adoptable. These aren't the only Plays. They're the first Plays, proven starting points for teams beginning their WISER practice.

As practitioners gain experience, they create domain-specific Plays. After one implementation, you have anecdotes. After two, hypotheses. After three, patterns worth sharing. Chapter 19 teaches you how to develop and validate your own Plays.

The Playbook

A Playbook is the memory that survives the chaos.

A Playbook adapts. Traditional plans execute once and collect dust. A Playbook absorbs every outcome, every adjustment, every hard-won insight. Run a Play, see what happens, update the Playbook. What worked becomes doctrine. What failed becomes warning. The Playbook guiding your next decision carries the memory of every previous one.

It captures the current state, objectives, and boundaries in a single place, preventing the insights that drove success in Solve from fading before they can be scaled in Expand. Every decision builds on previous learning rather than starting from scratch.

By making risks and constraints explicit, a Playbook prevents the organizational amnesia that kills momentum.

The principles at work: Individuals First. Deception Destroys Trust. Discovery Before Disruption. AI Inherits Messiness. (Individuals First governs the Playbook specifically; as a cross-cutting principle about protecting individual agency, it applies throughout the method rather than to a single Canon.)

Positions

Principle-driven tensions don't resolve themselves.

Someone must own the decision when the team can't agree. Someone must advocate for users when builders optimize for elegance. Someone must challenge assumptions before they become expensive failures.

Positions assign critical accountabilities to specific people. The Starter Plays define seven Positions that manage nine inherent tensions in AI implementation: Authority, Empathy, Translation, Context, Curiosity, Execution, Safety, Stewardship, Integrity.

Plays define how to fill these Positions for your team size and domain. One person covers multiple Positions on small teams. Larger organizations distribute them across specialists. The accountability matters more than the org chart.

The Outcome

Teams operating on WISER gain four capabilities they lacked before:

Continuous evolution. Rebuild systems while they run. No more waiting for the mythical rewrite that never comes.

Systematic risk burn-down. Identify and reduce risk through evidence, not avoidance. Each iteration makes the next one safer.

Living documentation. A Playbook that prevents organizational amnesia. Decisions survive personnel changes.

Clear ownership. Every decision has a person accountable, even as AI scales. No more "the algorithm did it."

The next five chapters show you what this looks like in practice.

We'll follow Wingman Services through all five Canons. They spent two years chasing AI scheduling solutions that vendors promised would transform their operations. What they actually needed was AI in a place nobody was selling.

You'll watch them discover the real problem, test their assumptions, build something that works, scale it across locations, and grow its autonomy over time.

The method becomes visible when you see it in action.

PART III

Part III: The Case Study

The method makes sense in the abstract. Canons, Playbook, Positions. But abstraction doesn't build capability. Practice does.

The next five chapters follow a single company through all five Canons. Wingman Services is a composite: the patterns come from hundreds of real implementations we've led or advised. We constructed this company to protect proprietary details while preserving the texture of how WISER actually works. The friction is real. The discoveries are real. The failures are real.

Wingman Services has been running for twenty years. Eight locations across three states. Three hundred employees. HVAC, plumbing, electrical, handyman. The kind of company that shows up when your furnace dies in February or your toilet backs up on Thanksgiving. The CEO founded it after leaving the Navy, where he'd spent a decade as a fighter pilot. The culture came with him. Meetings are briefings, post-mortems are debriefs, and nobody at Wingman blinks when someone calls an unidentified problem a bogey.

We join the Wingman team in their first WISER meeting. They've been wrestling with the same operational problem for four years.

PART III

Chapter 5: Witness

The **Sponsor** has been COO for eight years. He came up through operations, which means he knows the business. It also means he's been burned by promises that didn't deliver.

He's standing at a whiteboard covered in vendor logos. Three different companies have pitched him "AI-powered scheduling" over the past two years. Demos that dazzled. Pilots that fizzled. Features that sounded revolutionary until his dispatchers tried to use them.

"I'm done with AI demos," the **Sponsor** says. "Every vendor promises their AI will optimize our dispatch. None of them have delivered anything my team actually uses. But I know there's something here. Competitors are pulling ahead. I just can't figure out where AI actually helps versus where it's just marketing."

The **Guide** is an independent WISER Method consultant, brought in after the third vendor failed to deliver. She looks at the whiteboard. Then at the **Sponsor**.

"That's the right question," the **Guide** says. "But we can't answer it from vendor decks. Can we spend two weeks observing how your operations actually work? Not how the software says they work. How they actually work."

The **Sponsor**'s jaw tightens. "We already know how it works. Dispatchers get calls, assign techs, techs do jobs. The bottleneck is scheduling. That's where AI should help."

The **Architect** designed Wingman's customer self-serve portal two years ago. That project shipped on time and customers actually used it. That success earned her a seat on this one.

"If that's true," the **Architect** says, "two weeks of observation will confirm it. But if AI can help somewhere else, somewhere the vendors aren't looking, we'll find that too."

The **Sponsor** shakes his head. "We've already diagnosed this. Scheduling is the bottleneck. I don't need two more weeks of consultants watching my dispatchers work."

"You've diagnosed where vendors told you to look," the **Guide** says. "We're asking to look everywhere else."

The **Sponsor** glances at the whiteboard. Three vendors. Two years. Zero results. Two more weeks to find where AI actually fits doesn't seem unreasonable.

"Fine," he says. "Two weeks. Then I want to know exactly where AI can actually move the needle. Not where vendors want to sell it."

The **Sage** has been at Wingman for fifteen years, across three different roles. She started in dispatch, moved to field coordination, and now manages operations for the Denver region. She knows the history. She knows the people. When the team needs introductions, she opens doors.

"The lead dispatcher has been here as long as I have," the **Sage** tells the **Architect**. "She knows things the system doesn't. Watch her work. Don't judge it."

The **Architect** spends her first day observing the dispatcher.

Within the first hour, she notices something unexpected. The dispatcher never touches ServiceTitan for actual scheduling.

"Why don't you use the dispatch board?" the **Architect** asks.

The dispatcher pulls up her screen. ServiceTitan is open, but minimized. Next to it: a color-coded spreadsheet she's refined over fifteen years.

"That thing shows me where techs are," the dispatcher says. "But it doesn't know that one tech runs long on Tuesdays because his wife has chemo and he visits her at lunch. Doesn't know that certain customers always add three more items when we show up. Doesn't know that traffic on I-70 is worse after 3pm because of construction."

She gestures at her spreadsheet. Green cells for available techs. Yellow for lunch. Red for personal situations. Notes in every cell.

"This is how I schedule. The software is for billing."

The **Architect** takes notes on a system that exists in no process document. Fifteen years of pattern recognition, encoded in one spreadsheet that lives on one person's desktop.

> "A complex system that works is invariably found to have evolved from a simple system that worked. A complex system designed from scratch never works and cannot be patched up to make it work."
>
> John Gall, *Systemantics*

The dispatcher's spreadsheet is that simple system. It evolved because ServiceTitan couldn't handle reality. The **Architect** realizes they're not looking at a workaround. They're looking at the actual dispatch system.

The next day, the **Architect** and the **Scout** follow a single job from dispatch to invoice.

The **Scout** is a senior HVAC technician. Twelve years in the field. He's also on the WISER team because if he adopts a new system, other techs will follow. If he rejects it, they will too.

The job is straightforward: a residential furnace that won't ignite. The dispatcher assigns it via her spreadsheet. Then she updates ServiceTitan, because that's where billing looks for data. Two systems, already.

The **Scout** arrives at the customer's house, diagnoses the issue, replaces the igniter. Thirty minutes of actual work. Then he pulls out a paper form.

"What's that?" the **Architect** asks.

"Job ticket," the **Scout** says. "Parts used, time spent, customer notes. This is what billing needs."

He fills it out by hand, gets the customer's signature on a carbon copy, and drops the white copy in a folder on his passenger seat. End of day, he'll drop the folder in a basket at dispatch.

The **Architect** follows the paper. The next morning, a billing clerk pulls the form from the basket. She squints at the **Scout**'s handwriting, enters the data into the ERP, and generates an invoice.

The billing manager has agreed to let the team observe. "Please," she said. "Someone needs to see what we're dealing with."

The **Architect** watches the billing clerk process the morning's forms. Squinting at handwriting. Guessing at part numbers. Calling technicians who don't answer. By noon, three invoices are stuck waiting for clarification. Two have already gone out with errors that will require correction later.

The **Architect** maps the handoff points. Six handoffs across four systems, with paper in the middle of all of them.

Over the next week, the team quantifies what they've observed.

They shadow jobs across three locations. They interview billing clerks. They time handoffs. They count errors.

Midway through week two, the **Sponsor** finds the **Architect** in the break room.

"I've seen enough," he says. "Paper forms are the problem. Let's build a digital form and test it."

The **Architect** shakes her head. "We're not done observing."

"We've been watching for ten days. We know the problem is data capture. Why keep watching?"

"Because we know one part of the problem. We don't know how bad it is, or where else it shows up, or what we'll break if we change it." She pulls up her notes. "Yesterday I found out the billing manager has a second workaround for rush jobs. Day before that, I learned the dispatcher flags certain techs for overtime differently than the system tracks. Every day we find something new."

"So when do we stop finding things?"

"When we stop being surprised. Right now, every conversation reveals something we didn't know. When we're confirming what we already suspect, we're ready to test. Not before."

The **Sponsor** looks frustrated. "I'm paying for consultants to watch my people work."

"You're paying for us to understand your system before we change it. Every AI vendor who came in here skipped this step. They saw dispatch software and assumed scheduling was the problem. Two years, zero results."

The **Guide** walks in, catches the tension. "How many vendors showed you demos in the first week?"

"All of them," the **Sponsor** admits.

"And how many of those products are you still using?"

Silence.

"This is why we won't build the wrong thing," the **Guide** says. "This is the Witness Canon. We don't prescribe until we understand."

The **Sponsor** exhales. "Fine. Finish the week. But I want numbers. Not just stories."

The numbers are worse than expected.

Paper form completion happens at every job. Five to ten minutes per job. Twenty-five percent of entries are illegible or incomplete.

Entering paper data into the ERP takes eight minutes per job. Fifteen percent data entry errors on top of the form errors.

Billing clarification calls happen on nearly half of jobs. Twelve minutes per call. Blocks the invoice by at least twenty-four hours.

Invoice corrections are required on more than a third of invoices. Twenty minutes per correction. Customer trust erodes with each one.

When the friction compounds, it adds up to six hours per day per location. Across eight locations, that's forty-eight hours of daily waste. Plus the invoices that never get corrected, the customers who don't call back, the revenue that quietly disappears.

> "Every system is perfectly designed to get the results it gets."
> W. Edwards Deming

The system is producing exactly what it's designed to produce: paper handoffs that generate errors, errors that generate rework, rework that generates customer complaints. The scheduling software isn't the cause. It's barely involved.

Two weeks later, the team debriefs the **Sponsor**.

The **Guide** pulls up the friction map. Six handoff points, color-coded by error rate. The paper-to-billing transition is dark red.

"The scheduling software isn't the problem," the **Guide** says.

The **Sponsor** leans forward. "Then what is?"

The **Architect** points to the friction map. "The handoff. Technicians capture job data on paper. That paper becomes the source of truth for everything downstream. And paper can't be searched, verified, or automated."

The billing manager, who's sitting in, nods. "I've been saying this for years. We spend more time fixing invoices than sending them."

The **Sentinel** is Wingman's finance director. She's been tracking the financial impact. "If data capture is really the problem, this is significant. I've been watching margin leakage from billing errors for three years. It's not small."

The **Sponsor** pauses. This isn't the answer he expected. Every vendor demo focused on scheduling. Every pitch deck optimized dispatch. And scheduling was never the problem.

"So where does AI actually help?"

"Here," the **Architect** says, pointing to the paper-to-billing handoff. "Technicians capture job data by hand. That data has to flow to billing. Right now, humans transcribe it. Humans make errors. Humans call for clarification. Nearly 40% error rate at invoice."

The **Sponsor** stares at the friction map. "So the AI opportunity was never scheduling. It was data capture."

"The vendors were selling where they had products," the **Guide** says. "We found where you have problems."

The **Sponsor** considers this. Two years of demos that promised to optimize scheduling. Two weeks of observation that found where AI could actually help.

"Okay," he says. "What do we need to test to prove this?"

The Playbook gets its first entries.

Stakeholder map: The dispatcher (scheduling), the technicians (data capture), the billing clerks (transcription), the billing manager (corrections). Each with different pain points. Each with different definitions of "the problem."

Actual workflow: Dispatch happens in a spreadsheet. Paper forms carry data between field and office. Billing happens in the ERP. ServiceTitan is a passthrough for compliance, not operations.

Friction map: Six handoffs. Four systems. Paper in the middle. 38% error rate at invoice.

Baseline metrics: 6 hours/day/location in rework. 38% invoice correction rate. 3-day billing reconciliation.

Key assumptions to test: Can we improve data capture without changing technician behavior? Would a digital form help, or would technicians reject it? Can AI handle the transcription step?

The **Architect** captures all of it. "The Playbook holds everything," she says. "Decisions, constraints, what we've learned. When someone asks why we built it this way six months from now, the answer is here."

The **Sponsor** walks the team to the door.

"Not one vendor looked at the paper forms," he says. "They all wanted to optimize scheduling because that's where their product fit."

"Vendors sell what they have," the **Guide** says. "They're not incentivized to find where AI actually helps. WISER starts with observation because AI isn't magic. It needs to understand the real system, not the documented one."

The **Sponsor** nods. "I kept looking where everyone told me to look. Should have watched my own operation sooner."

> "Before you disturb the system in any way, watch how it behaves. If you don't understand the natural rhythms, you will never be able to harmonize with them."
>
> Donella Meadows, *Dancing With Systems*

The team watched. Now they understand. The dispatcher's spreadsheet isn't a workaround. It's the only thing that works. The paper forms aren't legacy. They're the load-bearing wall. The billing errors aren't operator failure. They're system design.

The teams that skip Witness are usually the ones who say "we already know the problem." They're often wrong. The understanding they've built over years may not survive contact with AI's different way of working. Observation reveals what assumptions have been hiding.

The next step is testing whether their diagnosis is right. Cheap experiments, each building on the last. If data capture is really the problem, they'll prove it. If something else is hiding underneath, they'll find that too.

The **Sponsor** has one more question.

"How long until we know if AI can actually solve this?"

"A few weeks," the **Architect** says. "We need to test a few things. Scheduling visibility first, just to rule it out. Then data capture approaches. Each experiment will inform the next."

"And if we're wrong about the handoff?"

"Then we find out before we build the wrong thing. Better than finding out after you've bought another AI product that doesn't fit."

The **Sponsor** looks at the friction map one more time. Every vendor had pointed at scheduling. None had pointed here.

"Okay," he says. "Run the experiments."

Chapter 6: Interrogate

Two days after the Witness presentation, the team reconvenes in Wingman's conference room. The friction map is still taped to the wall. Someone has added sticky notes with questions: "How do we know this is real?" and "What if we're wrong?"

Good questions. The team has a hypothesis, not proof. Data capture causes errors, not scheduling. But hypotheses are cheap. Building on wrong hypotheses is expensive.

The **Guide** stands at the whiteboard. Three items:

1. Technicians need better scheduling visibility
2. Digital data capture is better than paper
3. AI can eliminate transcription errors

"We're testing all three," she says. "Cheap experiments. Days, not weeks."

The **Sponsor** frowns. "We're still testing scheduling visibility? I thought we just proved scheduling wasn't the problem."

"We proved it wasn't where the errors come from," the **Architect** says. "But you've believed it was the problem for two years. So have your dispatchers. If we skip testing it, you'll always wonder if we missed something."

The **Sage**, sitting in the corner with a cup of coffee, speaks up. "I've been here long enough to know that people sometimes complain about the wrong thing when they can't name the right thing. Test it. Prove it. Then nobody can second-guess us later."

The **Sponsor** nods slowly. "Fair. Rule it out."

The **Architect** drags a whiteboard into the break room on Monday morning. She spends an hour transcribing the dispatcher's spreadsheet: technician names in rows, time slots in columns, job assignments in each cell. Color-coded sticky notes for job status. It looks professional. It looks like the visibility dashboard every vendor tried to sell.

By 7:30, technicians stream through for coffee. A few glance at the board. One guy squints, finds his name, nods, and walks out. Nobody stops to study it.

Day two. Same pattern. Coffee, glance, gone.

Day three. The **Architect** catches the **Scout** in the break room.

"Is the board useful?"

He looks at it like he's noticing it for the first time. "I look at my phone. Dispatch texts me my next job. Why would I come back here to check a board?"

She finds the **Sponsor** that afternoon. "Have you looked at the scheduling board?"

"The what?" He pauses. "Oh. Honestly, no. I assumed if something was wrong, someone would call me."

By day seven, the whiteboard has become a community bulletin board. Someone taped up a flyer for a Fantasy Football league. A birthday card for someone named Mike. The schedule is still there, but nobody's updated it since Wednesday.

The experiment is over. The **Architect** peels off the sticky notes and returns the whiteboard to the conference room.

The **Sage** is waiting. "No joy?"

"Not once."

She sips her coffee. "You just saved them from buying a $40,000 dashboard that would've ended up the same way."

The **Smith** has been building something. He's proud of it.

Two days of work: a mobile web form that mirrors the paper job ticket. Job type dropdown. Parts inventory with autocomplete. Time entry. Customer notes. Submit button that shoots the data straight to a Google Sheet for billing.

"No more paper," he says, demoing it to the team. "Technician fills this out at the truck. Data goes directly to billing. No transcription. No handwriting. No clarification calls."

The **Scout** looks at the phone screen. As the team's field validator, he's the natural first tester. "I'll run it through a week of real jobs."

Day one. The **Scout** finishes an HVAC repair, pulls out his phone, and opens the form. He taps through screens. Selects the job type. Scrolls through the parts dropdown looking for "capacitor." Doesn't find it. Types it manually. Moves to time entry. Fat-fingers the minutes. Has to scroll back and fix it.

Four minutes later, he hits submit.

He used to fill out the paper form in ninety seconds.

Day three. Another technician texts the **Architect**: "This thing is more work than paper. And I can't load it in the truck because I lose signal in half my neighborhoods."

Day five. The **Smith** checks the analytics. Four technicians started the week: the **Scout** plus three volunteers from the field crew. Usage has dropped to 20%. The **Scout** is the only one still submitting forms, and his submissions are getting later and later in the day, bunched together like he's catching up at the end rather than doing them at each job.

The debrief is quiet. The **Smith** stares at the usage chart on the wall.

"We built exactly what they asked for," he says. "Mobile. Digital. Faster path to billing."

The **Architect** shakes her head. "We built what we thought they needed. They needed something faster than paper. We gave them something slower." She pauses. "That's why WISER runs cheap experiments first. A week of testing cost us a few days of the **Smith**'s time. A full rollout would have cost us months of trust."

The **Scout**, sitting in the corner, shrugs. "If you want me to capture data in the field, it has to be faster than paper. Not slower. Not the same. Faster."

The **Smith** doesn't say anything. He closes his laptop.

The **Sage** catches him in the hallway afterward.

"You okay?"

"No." He exhales. "But I get it now. I built what I thought was elegant. They needed what was fast."

She nods. "That's worth knowing."

The team pivots.

If technicians won't abandon paper, maybe AI can meet them where they are.

The setup is simple: technicians keep filling out paper forms. They like paper. Paper is fast. But instead of dropping forms in a basket at dispatch, they photograph each completed form with their phone. The photo goes to a shared folder. AI extracts the structured data, fills a Google Sheet for billing.

When the AI can't read something, it texts the technician: "Can't read 'parts used' on job #4721. Reply with the part name."

The **Scout** runs the field test himself. As a working technician on the WISER team, he can validate whether this works in real conditions. "At least I'm not changing what I do in the field."

Week one. Forty job tickets photographed.

The **Sentinel** pulls the billing data at week's end. She's been tracking invoice errors for three years. She knows exactly what 38% error rate looks like in dollars.

"AI correctly extracted 85% of fields on first pass," she says. "Twelve percent required one clarification text. Three percent needed manual review."

The billing manager, who's joined the meeting, looks up from her notes. "We went from 38% invoice errors to 9%. And we sent invoices same-day instead of waiting for clarification calls."

The **Sponsor** is on his feet before anyone else speaks. "This is it. Photo plus AI. Let's roll it out."

The **Guide** holds up a hand. "Not yet."

"Why not? We have a working solution. We have the numbers. Why keep experimenting?"

"Because we have a working experiment," the **Architect** says. "Not a production system. And we have a hypothesis we haven't tested."

"What hypothesis? The photo thing works."

The **Smith** speaks up. "Photo of paper works. But technicians are still filling out paper forms. That's five to ten minutes per job. What if we could eliminate the paper entirely?"

The **Sponsor** frowns. "You just said the photo system works. Why change it?"

"Because 'works' isn't the same as 'best,'" the **Guide** says. "Photo-AI is our proven fallback. But if photo-voice works, we're not just eliminating transcription errors. We're eliminating the form itself. That's a different order of magnitude. The Interrogate Canon says test before you commit. We haven't finished testing."

The **Sentinel** pulls up a spreadsheet. "Photo-AI saves us $45K to $50K monthly in billing accuracy. If photo-voice eliminates forms, we save another 8 to 10 hours per day in technician time. That's labor they can redirect to revenue-generating jobs."

The **Sponsor** looks at the numbers. Then at the **Guide**.

"How long to test the photo-voice hypothesis?"

"A week to build a prototype. Another week to field test. If it fails, we've lost two weeks. If it works, we've found something much bigger."

"And if we just roll out photo-AI now?"

"Then we've locked in a solution without knowing if a better one was two weeks away. And changing systems mid-rollout is expensive."

The **Sponsor** sits back down. "Fine. Test it. But photo-AI is our floor. Whatever you build has to beat it, or we go with what works."

"Good," the **Guide** says.

Week two. Eight technicians join the photo-AI experiment while the **Smith** builds the photo-voice prototype. The AI learns individual handwriting patterns. Accuracy improves to 92%. Clarification texts drop to 5%. Error rate holds at 9%.

The **Sentinel** updates the financial analysis. "Photo-AI is confirmed at $45,000 to $50,000 per month in recovered billing."

The **Smith** is watching the numbers. His digital form failed. Photo-AI worked. The difference: photo-AI didn't ask technicians to change.

"We eliminated the transcription step," he says. "But we didn't eliminate the paper."

The **Architect** nods. "That's the next question."

The closing debrief. Three experiments. Three weeks.

"Scheduling visibility is dead," the **Guide** says, crossing it off the whiteboard. "Digital forms are worse than paper. Photo plus AI works."

"But we're still using paper," the **Scout** says. He's been in every debrief, testing every iteration. "I don't mind photographing it. That's easy. But you said something about eliminating paper entirely."

The **Architect** looks at him. "What if you didn't fill out a form at all? Photo of the completed job. Voice note describing what you did. AI extracts everything."

The **Scout** considers this. He's the one who'll have to sell it to the other technicians if it works. "Photo of the job. Voice note. That's it?"

"That's the hypothesis."

"How long to build it?"

The **Smith** has been quiet since the digital form failure. Now he speaks up. "A week for a rough prototype. Maybe less."

"And if it doesn't work?"

"Then we've learned something. Photo plus AI transcription already works. That's our fallback."

The **Sponsor** looks at the experiment results on the wall. Three weeks ago he was convinced AI needed to optimize scheduling. Now he's looking at a path to eliminate paper forms entirely.

The **Sponsor** stands up. "Build the prototype."

The Playbook gets updated that evening.

Experiment 1: Scheduling visibility. Nobody used it. Hypothesis invalidated.

Experiment 2: Digital forms. Slower than paper because tapping through screens took longer than scribbling. Technicians abandoned it. Hypothesis partially confirmed: digital capture could work, but not with forms.

Experiment 3: Photo + AI transcription. 38% error rate dropped to 9%. Technicians didn't change behavior. Hypothesis confirmed.

Next hypothesis: Photo + voice could eliminate paper entirely.

Key insight, added by the **Sage**: Meet users where they are. The **Scout** won't change how he works. But he'll photograph his work if it takes less time than dropping forms in a basket.

Risk logged by the **Sentinel**: AI transcription is only as good as photo quality. Low-light conditions and damaged forms may degrade accuracy. Monitor edge cases in production.

The **Smith** stays late, setting up the development environment for the prototype. The **Architect** finds him still there at 8pm.

"You okay?"

"The digital form was good work," he says. "Clean code. Good UX. It just solved the wrong problem."

"That's what experiments are for. Fail cheap, learn fast."

He nods. "The photo-voice thing. I think it'll work. But I'm not going to assume anymore. We'll test it with real technicians before I write a single line of production code."

The **Architect** nods.

She turns off the lights on her way out. The **Smith**'s laptop glows in the dark.

Tomorrow they build.

PART III

Chapter 7: Solve

The **Smith**'s laptop is open to a blank code editor. It's 6am. He's been here since five.

Photo-voice. No paper at all. Technician photographs the completed job, records a voice note, AI extracts everything billing needs. That's the hypothesis. Now he has to build it.

The constraint is burned into his brain: fifteen seconds. That's the max the **Scout** will tolerate in a customer's driveway. Ten is better. Eight is the target.

He starts typing.

By Thursday, there's a working prototype. Rough, but functional. The **Smith** demos it to the team in the conference room.

"Technician opens the app. Takes a photo of the completed work. Taps record, says what they did. Taps stop. AI extracts job type, parts used, time spent, location. Technician confirms or corrects. Data goes to billing."

The **Architect** watches the screen. "How long?"

"In testing? About twelve seconds. But that's me in a quiet office. Field conditions will be different."

The **Scout** leans forward. "Let's find out."

They drive to a job site. Residential HVAC, furnace replacement. The **Scout**'s already finished the work. The customer is inside, waiting for the paperwork.

The **Scout** pulls out his phone, opens the app, and photographs the new furnace. Then he taps record.

"Replaced Carrier furnace, model 59SC5A, gas valve and igniter. Customer approved the work. Took about two hours including cleanup."

He taps stop. The screen shows a processing spinner. Three seconds. Then structured data appears: job type, parts, duration, completion status.

The **Scout** squints at the screen. "Parts are wrong. It says 'gas valve.' I replaced the whole unit."

"Tap to correct," the **Smith** says.

One tap. A list of options. The **Scout** selects "Full unit replacement." Total time from photo to confirmation: fourteen seconds.

"Not bad," the **Scout** says. "But that was quiet. Let me try it on a noisy one."

The next job is a commercial rooftop unit. The compressor is running. Wind is blowing across the roof. The **Scout** finishes the repair and pulls out his phone.

He takes the photo. Taps record. Starts speaking.

The AI returns garbage. "JOB TYPE: Unknown. PARTS: [unintelligible]. DURATION: Error."

The **Scout** shows the screen to the **Smith**. "Flameout. It can't hear me over the compressor."

The **Smith**'s face falls. "The voice recognition worked perfectly in the office."

> "We are searching for the 'Jagged Frontier.' AI is brilliant at things you expect to be hard and fails at things you expect to be easy. You cannot predict the boundary; you have to walk it."
>
> Ethan Mollick, *One Useful Thing*

The AI had no trouble parsing complex part names and technical vocabulary. It choked on background noise. The **Smith** expected the hard part to be understanding "59SC5A." The actual hard part was wind.

Back in the conference room, the team debugs.

"We need noise cancellation," the **Smith** says. "Or better microphones. Or..."

"Or a fallback," the **Architect** interrupts. "What if the technician can't use voice? What do they do?"

"They take a photo of the paper form," the **Sage** says. "Like we did in the last experiment. It worked."

The **Scout** nods. "If I can't speak, I can scribble. Thirty seconds with paper, then photograph it. Still faster than driving back to drop forms in a basket."

The **Smith** writes it down. "Primary path: photo-voice. Fallback path: photo of paper. The AI handles both."

"And we train the voice model on real field audio," the **Architect** adds. "Not office recordings. Compressors. Truck engines. Wind. Dogs barking."

The **Guide** opens the Playbook on her laptop. "Decision logged: dual-path capture. Primary is voice. Fallback is paper photo. Rationale: voice fails in noisy conditions, but technicians still need a path that works."

She adds it to the decision log, timestamps it, notes who was in the room.

The **Smith** looks tired. "That's another week of work."

"Better than six months of patching a system that doesn't work in the field," the **Guide** says.

Week two. The **Smith**'s team collects field audio from a dozen job sites. Rooftop units. Basements with running dehumidifiers. Driveways with lawnmowers next door. They retrain the voice model.

The **Scout** runs tests. Results improve. Voice recognition accuracy climbs from 60% to 85% in noisy conditions. The fallback to paper photo catches the rest.

Every failed transcription becomes training data. The AI learns the **Scout**'s voice, his part vocabulary, his phrasing. It starts anticipating patterns: "furnace" usually means HVAC, "disposal" usually means plumbing, "panel" usually means electrical.

By end of week two, the **Scout** texts the **Smith**: "Eight seconds on the last five jobs. Voice worked every time. Even next to a running truck."

Week three. Denver pilot.

Twelve technicians. Four weeks. But there's a critical piece the team almost missed.

"Who validates the AI's output?" the **Architect** asks. "During pilot, we can't just trust it."

The billing manager volunteers. She's been reviewing invoice errors for three years. She knows what correct job data looks like. She knows when a part name is wrong, when a duration doesn't match the job type, when something smells off.

"I'll review every submission," she says. "Every photo, every voice transcription, every extracted field. If the AI hallucinates, I'll catch it."

The **Guide** adds it to the Playbook: Human-in-the-loop validation. Every submission reviewed during pilot phase. Errors documented with screenshots. Patterns fed back to the **Smith** for model improvement.

"AI First Principles says humans direct AI, not the reverse," the **Architect** adds. "During pilot, that means every output gets human eyes. We'll loosen the grip later, once we've earned it."

The **Scout** recruits technicians personally. "I was skeptical too," he tells them in the break room. "But it's faster than paper. Try it for a week. If you hate it, go back to the forms."

Week one: 60% adoption. The holdouts don't trust the sync. The **Scout** converts them one by one.

Tuesday afternoon, the billing manager is reviewing submissions when she stops cold. A routine water heater replacement. The technician's voice note was clear: "Replaced 50-gallon Rheem, gas." The AI extracted: "Duration: 8 hours."

She pulls up the photo. Standard tank swap. Two hours, maybe three with complications. Eight hours is impossible.

She screenshots the submission, adds it to her error log, and walks to the **Smith**'s desk.

"Look at this."

The **Smith** stares at the screen. "Eight hours? Where did that come from?"

"The technician said 'took about two hours.' The AI heard 'eight.' Background noise, maybe. Or it's pattern-matching wrong."

"Can you flag all the duration errors this week? I need to see if there's a pattern."

By Friday, she's flagged six submissions. Two wrong part names. Three unclear photos. And the duration hallucination, which turns out to be a pattern: the model struggles with numbers spoken in certain accents.

"Hallucination rate is about 12%," she tells the team at standup. "But it's not random. It's struggling with plumbing vocabulary and some speech patterns."

The **Smith** retrains over the weekend. Week two: hallucinations drop to 4%. Adoption climbs to 85%. The holdouts are watching their colleagues finish faster.

Week three: 100% adoption. The billing manager catches 3 errors out of 156 submissions.

"I'm building a hallucination log," she says. "Part names it confuses. Phrases it misinterprets. When we roll out to Austin, the next validator will have a checklist."

The **Sentinel** pulls the billing data.

Invoice errors: 9%. Down from 38%.

Clarification calls: near zero.

Billing reconciliation: same-day. Down from three days.

Week four confirms the pattern. Error rate holds at 9%. Technicians average eleven seconds per job capture. Paper form usage drops 80%. The remaining 20% is the fallback path for noisy environments.

The **Sentinel** runs the financial analysis.

"Denver alone: $47,000 in recovered billing this month. That's revenue that would have been disputed, delayed, or written off."

The **Sponsor** stares at the number. "Forty-seven thousand. One location. One month."

"Conservative estimate. I'm not counting technician time savings. Or reduced customer disputes. Or the billing clerks who aren't squinting at handwriting anymore."

The **Smith** is quiet. Six weeks ago his digital form failed because it was slower than paper. Now the photo-voice system is being adopted voluntarily. The difference: they built what technicians would actually use, not what looked good in a demo.

"Extrapolate to all eight locations," the **Sponsor** says.

"Rough math: if other locations perform at even half Denver's rate, we're looking at $2.5 million annually in recovered billing. Plus operational savings I haven't quantified yet."

The **Sponsor** stands up. "I need to present this to the executive team."

The executive presentation. The **Sponsor** at the whiteboard, the CEO and CFO watching.

"Denver pilot. Four weeks. Invoice errors down from 38% to 9%. Same-day billing instead of three-day reconciliation. $47,000 recovered in one month."

The CFO leans forward. "Extrapolated?"

"Conservatively, $2.5 to $3 million annually across all locations. Plus operational efficiencies we haven't fully measured."

The CEO leans forward. "When can we roll this out everywhere?"

The **Guide** speaks up. "We recommend a sequenced approach. Austin next, because it's similar to Denver. Then Phoenix, which is commercial-only. Different context, different risk profile." She pulls up the Playbook. "That's the Expand Canon. Prove it works in one context before you scale to another."

The CEO considers this. "I understand caution. But we have momentum. What's the risk of moving faster?"

The **Architect** answers. "Because we haven't proven it works in commercial contexts. Denver is 90% residential. Commercial jobs take multiple days. Our solution assumes same-day completion."

The CEO leans back. "Here's what I keep thinking about. If AI can fix billing, why stop there? ServiceTitan is a mess. The ERP is a mess. The dispatcher's spreadsheet is a mess. What if we rip out the whole thing and rebuild with AI from scratch? One integrated system."

The room goes quiet.

The CFO looks interested. "A full ERP replacement?"

"I'm not saying tomorrow. But we've got proof AI works. At some point, patching old systems has diminishing returns. When do we think bigger?"

The **Guide** glances at the **Sponsor**. This is the moment.

The **Sponsor** stands up. He walks to the whiteboard where the pilot results are still displayed.

"I get the impulse," he says. "I've wanted to blow up our systems for years. But here's what I've learned watching this team work."

He draws a box. Labels it "ERP."

"This is our system. It's ugly. It's held together with duct tape and workarounds. But it runs. Every day. It handles dispatch, billing, inventory, payroll. If we rip it out, we're not just replacing software. We're replacing every process that touches it. Every person who knows how it works. Every integration we've forgotten about."

He draws smaller boxes inside the big one. "Data capture. Dispatch. Billing. Inventory. Scheduling. Customer records."

"What this team just proved is that you can replace pieces. One component at a time. We didn't touch dispatch. We didn't touch inventory. We replaced one handoff: technician to billing. And we validated it works before we moved on."

The CEO frowns. "That sounds slower than I'd like."

"It's safer. And it's actually faster in the long run. Because when you rip out everything at once, you spend the next two years debugging integrations you didn't know existed. I've seen it happen. Twice. At companies that don't exist anymore."

The **Architect** speaks up. "The modular approach has another advantage. Each component we replace teaches us something. Data capture taught us about field conditions, about voice recognition limits, about human validation. When we tackle dispatch or inventory, we'll know how to run the process. We'll have people who've done it before."

The CEO looks at the **Sponsor**. "You've been through ERP replacements before. What's your take?"

"I think clean slate is how you end up with a $10 million project that's three years late and still doesn't work. Piece by piece is how you end up with a system that actually runs your business. We expand this solution region by region. We refine it until it's hardened for Wingman. Then we pick the next component. Dispatch, maybe. Or inventory. Same process, same team, same discipline."

He taps the whiteboard. "We iterate so we don't rip out something that's holding the machine together."

The CEO is quiet for a moment. Then: "Fine. Austin, then Phoenix. But I want weekly updates. And when this is proven across all locations, we talk about what's next. Dispatch. Inventory. Whatever makes sense."

The **Sponsor** nods. "Let's prove we've got a flight plan before we go supersonic."

The CEO raised an eyebrow. "Now you're speaking my language."

"Hard not to around here."

The Playbook captures the build.

User flow: How data moves from technician to billing, including the human validator and feedback loop.

System map: Which components talk to which, from field app to ERP.

Results: 38% errors down to 9%. Same-day billing. Eleven seconds per capture.

Key insights: Build for real conditions, not demo conditions. Human-in-the-loop isn't optional during pilot.

Risk flagged: Commercial contexts may break assumptions. Test before scaling.

Late Friday. The **Smith** is packing up when the **Scout** stops by.

"Hey. Thanks for sticking with it."

The **Smith** looks up. "The digital form failure still bugs me."

"It shouldn't. That failure is why this worked. You learned what not to build."

"I learned that I was building for myself, not for you."

The **Scout** shrugs. "Now you know. That's worth something."

He heads for the door, then turns back.

"Eight seconds. In a customer's driveway. With a truck running next to me." He grins. "I can live with eight seconds."

The **Smith** watches him go.

Tomorrow they start on Austin.

Chapter 8: Expand

Austin proved it wasn't a fluke. Same approach, same results, different people. Two locations validated the solution worked beyond the team that built it.

Now the CEO wants all eight locations live in sixty days.

"Help me understand something," he says, leaning forward in his chair. "Denver works. Austin works. At what point does caution become its own risk? Our competitors aren't waiting."

The **Guide** glances at the **Architect**. They've had this conversation privately. Now they're having it in front of the executive team.

"Because Denver and Austin are both residential markets," the **Architect** says. "Phoenix is 100% commercial. Different contexts may break the solution."

"Walk me through the risk. It's still HVAC. Still plumbing. Still technicians capturing job data. What breaks?"

"We're about to find out."

> "The fact that a solution worked once is often a matter of local luck, not universal truth."
> Nassim Nicholas Taleb

Austin went smoothly. That was the point. The team chose Austin first because its profile matched Denver: 85% residential, 15% commercial. Similar customer mix. Similar job patterns. Similar technician workflows.

They deployed with Denver's configuration. Adoption started at 70%, climbed to 95% by week three. Some techs preferred different voice phrasing, and the AI adapted. By week four, error rate was at 9% with same-day billing.

The **Sponsor** presented the results to the executive team.

"Austin validates the approach. Two locations. Consistent results. We're ready to expand."

The CEO nodded. "Then let's go. All remaining locations. Sixty days."

The **Guide** spoke up. "We recommend Phoenix next. Alone."

"Why Phoenix alone? Why not Phoenix plus three others?"

"Because Phoenix is our first commercial-only location. The context is different enough that we need to test it separately."

The CEO leaned forward. "Walk me through that. What specifically might break?"

"Commercial jobs span multiple days," the **Architect** said. "Our solution assumes same-day completion. We don't know if that assumption holds."

The CEO considered this. "So if Phoenix works, what would we need to see to feel confident about parallel rollout next time? I'm trying to understand what 'proven' looks like."

The **Guide** exchanged a glance with the **Architect**. The CEO was asking the right question.

"Two things," the **Guide** said. "First, the context-specific issues surface and get resolved. Second, we document what made Phoenix different so we know which other locations need similar treatment."

The **Sponsor** spoke up. "I pushed for speed at the start of this project. I learned better. If the team says Phoenix needs to be tested alone, there's a reason."

The CEO nodded slowly. "Okay. Phoenix first. But document what you learn. If we can parallelize after this, I want to know how."

"That's exactly what we're hoping to prove," the **Guide** said.

Phoenix's profile: 100% commercial. Service contracts with property management companies. Multi-building maintenance agreements. Jobs that span days, not hours.

The team deployed V3 with Austin's configuration. Problems were immediate.

The first call came from the Phoenix dispatcher on day two.

"The system is generating invoices for jobs that aren't done."

The **Architect** pulled up the logs. A commercial HVAC job at a strip mall. Three-day project: diagnosis on Monday, parts ordered Tuesday, installation Wednesday. The technician captured "diagnosis complete" on Monday afternoon. V3 generated an invoice.

"It thinks the job is done," the **Smith** said, staring at the screen.

"Because every job in Denver and Austin was done in one day," the **Architect** said. "The AI learned that 'capture' means 'complete.' In commercial, capture happens at every phase."

By the end of week one, the team was taking flak from all directions. Customers were receiving multiple invoices for the same project. Billing couldn't reconcile payments because the system showed three "completed" jobs instead of one in-progress project.

Error rate climbed to 55%. Worse than before the system existed.

The **Guide** made the call. "We're pulling the pilot. Phoenix goes back to paper until we fix this."

The dispatcher exhaled with relief. "Thank God. I've been bypassing the system for two days anyway."

Within an hour, Phoenix was back on manual processes. Paper forms, basket drops, billing clerks squinting at handwriting. The old way. Not great, but stable.

The CEO called an emergency meeting.

"Phoenix is back on paper. We've lost two weeks of progress."

The **Sponsor** shook his head. "We haven't lost anything. We learned that V3 doesn't work in commercial contexts. That's valuable information."

"I understand that. But Phoenix is back on paper. That's not progress."

"We're running Phoenix the way it ran before we started," the **Guide** said. "The baseline wasn't broken. We tried something that didn't work. We pulled it before it caused real damage. Phoenix is stable."

The CEO frowned. "Stable isn't progress."

"Stable is the foundation for progress. Phoenix isn't bleeding money right now. It's operating the way it always operated. We have time to build the right solution instead of patching the wrong one."

The **Sponsor** leaned forward. "If we'd rolled out to all eight locations at once, we'd have eight locations back on paper right now. Because we sequenced, we have one."

The **Guide** stepped in. "Phoenix isn't a failure. It's a discovery. One of the AI First Principles: AI learns from what it sees, not what you intend. We trained on residential patterns. Commercial has different patterns. The AI didn't fail. Our assumptions did."

The CEO looked at her. "So what do we do? Rebuild the whole system?"

"No. We adapt it. But we need to understand the commercial context first. That means going back to Interrogate."

"Interrogate? We already did that. In Denver."

"We did it for residential. Commercial has different patterns. We need to observe before we build."

The CEO exhaled. "How long?"

"Three days to understand the context. Then we'll know what to build."

The CEO looked at the **Sponsor**.

"Three days," the **Sponsor** said. "Then the team presents a plan."

The **Architect** flew to Phoenix that afternoon. The **Sage** joined via video, providing context from her years of working with commercial accounts.

"Commercial jobs have phases," the **Sage** explained. "Diagnosis. Parts ordering. Installation. Testing. Sometimes punch list items after that. Each phase can take a day or more."

The **Architect** shadowed a Phoenix technician through a three-day rooftop unit replacement. She watched him capture progress at each phase: photos of the existing unit, notes on parts needed, photos of the installation in progress, final completion capture.

"In residential, you show up and fix it," the technician said. "In commercial, you show up, assess it, order parts, come back, install, come back again to verify. Different rhythm."

The **Architect** mapped the differences:

Aspect	Residential (Denver/Austin)	Commercial (Phoenix)
Job duration	Same day	3-5 days typical
Capture points	One (at completion)	Multiple (each phase)
Billing trigger	Job complete	Milestone complete
Dispatcher need	Where's the tech now?	What's the job status?
Customer expectation	One invoice	One invoice at project end

The problem wasn't the AI. The problem was the assumptions baked into V3. Every design decision assumed same-day completion because that's what the team had observed in Denver.

Day three. The **Architect** presented findings to the team via video conference.

"V3 assumes one capture per job. Commercial needs multiple captures per job: in-progress, parts-waiting, customer-approval, complete. V3 triggers billing on capture. Commercial needs billing on milestone. V3 shows completion status. Dispatchers need in-progress status."

The **Smith** was already sketching on a whiteboard. "So we add job status. Dropdown: In-Progress, Parts-Waiting, Ready-for-Invoice, Complete."

The **Scout**, joining from Denver, spoke up. "Sounds like more taps."

"Two taps," the **Architect** said. "Status and photo. Voice optional for progress notes. Still under ten seconds."

"What about billing?" the **Sentinel** asked. "How do we prevent invoices from firing on every capture?"

"Milestone billing triggers," the **Smith** said. "Invoice only fires when status changes to Ready-for-Invoice. Everything else is progress tracking."

The **Guide** was updating the Playbook in real time. "I'm documenting this as a context-specific adaptation. V3 remains unchanged for residential. We're building V3.1 for commercial."

The CEO, who had been listening silently, spoke up. "How long?"

"Two weeks to build," the **Smith** said. "One week to test in Phoenix. One week to refine based on what we learn."

"Four weeks. Phoenix sits on paper for four more weeks."

"Phoenix is stable on paper," the **Guide** said. "They're not losing money. They're operating the way they operated before we started. Four weeks to build the right solution is faster than four months of patching the wrong one."

The CEO stood up and walked to the window. "This is exactly what I was afraid of. We had momentum. Now we're back to waiting."

The **Sponsor** leaned forward. "We're not waiting. We're building. The team identified the problem in three days. They have a solution design. Four weeks to execute is aggressive, not slow."

The CEO turned back. "I know I keep pushing. But I've watched what happens when we skip steps." He looked at the **Architect**. "What does the team need to move as fast as possible without breaking things again?"

"Exactly what we've proposed," the **Architect** said. "Two weeks to build, one to test, one to refine. We can cut corners, but we'll be back here in six weeks explaining why V3.1 failed too."

The CEO looked around the room. The **Guide**. The **Architect**. The **Smith**. The **Sponsor**.

"Four weeks," he said finally. "But I want weekly updates. And if this doesn't work, we're having a different conversation."

"Understood," the **Guide** said.

Week two. The **Smith** hits a wall.

He's been staring at the ERP API documentation for three hours. The same page. The same limitation. Looking for a workaround that doesn't exist.

The team is on a video call. The **Smith** shares his screen.

"The ERP can't handle the way we've built milestone billing. The integration needs rework."

His voice is flat. The **Architect** recognizes the tone. It's the same voice from Interrogate, when the digital form failed.

"V3 sends one invoice per job because residential is one capture, one invoice. V3.1 needs multiple captures with billing only on milestone. But the ERP treats every data push as a new invoice. There's no 'update existing job' in their API."

"So we need an intermediary layer," the **Architect** says. "Something to hold job state until billing."

"That's another two weeks. Minimum. Maybe more."

The **Smith** closes his eyes. "I should have checked this before I gave the timeline. I assumed. Again."

The **Guide** speaks carefully. "We promised four weeks. You're saying we need six."

"Maybe seven. I don't know what else I'll find." He opens his eyes. "I told the CEO four weeks because I wanted it to be four weeks. I wanted to be the guy who delivered on time after Phoenix blew up. That was ego. Not engineering."

The **Sentinel** pulls up the project timeline. "He's expecting deployment in two weeks."

"I know. And I have to tell him I was wrong."

The **Guide** calls the **Sponsor** before the weekly update. She explains the ERP problem. The intermediary layer. The revised timeline.

"Six to seven weeks instead of four," he says. "The CEO is going to lose his mind."

"Four weeks was what he wanted to hear. Six weeks was what he needed to hear. That's on us." She pauses. "AI First Principles says transparency earns trust. We have to be honest about what we don't know, not just what we do."

The **Sponsor** stares at the whiteboard where Phoenix's error rates are still written. 55%.

"So we tell him the truth?"

"We tell him the truth."

The weekly update. The CEO, the **Sponsor**, the **Guide**, the **Smith**.

The **Guide** presents the situation directly. No softening.

"We hit a technical constraint with the ERP. Their API doesn't support milestone billing. We need to build an intermediary layer. That adds two to three weeks to the Phoenix timeline."

The CEO's face hardens. "You told me four weeks."

"We did. We were wrong."

"Wrong how? You didn't know about the ERP?"

The **Smith** speaks up. "I assumed the integration would work the same way it worked for residential. It doesn't. Commercial billing is fundamentally different, and the ERP wasn't built for it."

"So you guessed."

"I estimated based on incomplete information. When I had complete information, the estimate changed."

The CEO is quiet for a moment.

"I appreciate the honesty." The CEO paused. "I'd rather hear hard truths early than easy promises that fall apart. But this is still hard to hear. Phoenix has been on paper for three weeks. Now you're telling me six more. Maybe seven."

The **Sponsor** stands up too.

"The team made a mistake. They gave you a timeline they hoped was true instead of one they knew was true. That's on them. But the process isn't broken. The process caught a constraint we didn't know existed. If we'd deployed V3.1 without the intermediary layer, Phoenix would have failed again. We'd be having this same conversation, except we'd also have angry customers and broken billing."

He turns back to the room. "So what's the actual timeline? Not the one you think I want to hear. The real one."

The **Smith** answers. "Six weeks from today. Two weeks to build the intermediary. Two weeks to integrate and test. Two weeks of buffer for whatever else we don't know yet."

"Six weeks."

"That's the timeline I can commit to. If it changes, I'll tell you before the weekly update, not during it."

The **Guide** adds: "We made an error in how we communicated. We gave you what you wanted to hear. Going forward, we give you what you need to hear. Even when it's not what you want."

The CEO looks at the **Sponsor**.

"You trust this team?"

"They made a communication mistake and owned it. No hiding. No excuses. Just the truth and a request for more time."

The CEO sits back down.

"Six weeks. But I want honest updates. If the timeline changes, I hear about it immediately. Not in a meeting."

"Understood," the **Guide** says.

V3.1 deployed to Phoenix in week five. The intermediary layer held job state until billing milestones. Status capture added two taps. The dispatcher dashboard showed real-time project progress.

The Phoenix dispatcher had been on paper for five weeks. She was skeptical.

"I spent two weeks cleaning up after your last version," she said. "The techs still don't trust the system. Neither do I."

"That's fair," the **Architect** said. "We broke it. We're asking for one more day to prove we fixed it."

The dispatcher crossed her arms. "One day. And if it generates another garbage invoice, we're done."

Day one: a multi-day commercial job. The technician captured "diagnosis complete." No invoice fired. Day two: "parts ordered." The dispatcher saw job status without calling anyone. Day three: a customer called asking for an update. The dispatcher pulled up the dashboard and told them exactly where things stood. Day four: "Ready-for-Invoice." One invoice generated for the complete project.

The dispatcher called the **Architect** that afternoon. "It works. This is what we needed."

By week seven, Phoenix error rate had dropped to 11%. Not Denver's 8%, but dramatically better than the 55% disaster. The commercial adaptation worked.

The **Guide** updated the Playbook:

Context Comparison Matrix:

Context	V3 (Residential)	V3.1 (Commercial)
Job duration	Same day	Multi-day
Capture model	Single capture	Multi-phase capture
Billing trigger	On capture	On milestone
Dispatcher view	Completion status	Progress status
Error rate target	< 10%	< 15%

Phoenix iteration documented: Initial deployment failed due to context mismatch. Three-day observation identified root cause. V3.1 developed and tested. Error rate recovered from 55% to 11%.

Key learning: Context isn't cosmetic. Same industry, same company, fundamentally different operating logic. Scaling means testing whether assumptions transfer, not assuming they do.

Five locations were now live. The remaining three rolled out over six weeks, two locations per wave, V3 for residential and V3.1 for commercial. Each wave was monitored for surprises.

There were none.

Week eighteen. All eight locations live.

The **Sentinel** presented the consolidated results:

Residential (V3):

- Invoice error rate: 9% (down from 38%)
- Same-day billing: 94% of jobs
- Technician capture time: 11 seconds average

Commercial (V3.1):

- Invoice error rate: 12% (down from 38%)
- Milestone billing accuracy: 96%
- Project status visibility: Real-time for all active jobs

Financial impact:

- Recovered billing: $247,000/month across all locations
- Extrapolated annual impact: $2.96 million

The **Scout** sent a message to the team chat: "Eight locations. Eight seconds. Still works."

The **Smith** replied: "Nine seconds in Phoenix. The commercial status dropdown adds a tap."

The **Scout**: "I can live with nine seconds."

The Playbook captures the expansion.

Context matrix: How residential and commercial differ in job duration, capture points, and billing triggers.

Phoenix iteration: What broke, why, and how V3.1 adapted.

Key insight: Local success doesn't guarantee universal success. Each context tests whether assumptions transfer.

The **Sponsor** walked the **Guide** to the parking lot after the final expansion meeting.

"A year ago, I would have pushed to roll out everywhere on day one," he said. "I would have blamed the team when Phoenix failed. I would have demanded they fix it faster."

"What changed?"

"I watched the process work. Denver proved the solution. Austin proved it wasn't a fluke. Phoenix proved we didn't know everything. The sequencing caught the problem before it spread."

He paused by his car, keys in hand. "You know what I keep thinking about? Nobody got laid off. We're not cutting the billing team. We're expanding what they can do. The clerks who used to chase data entry errors are becoming the people who catch what the AI misses. That's not efficiency theater. That's capability."

"That was a choice you made early. Not everyone makes it."

"It was the right choice. And watching it play out made everything else easier. The team trusted the process because they weren't afraid of it."

The **Guide** nodded. "That's the point of Expand. You're not just scaling the solution. You're scaling the learning. Each location teaches you something."

"Phoenix taught me that I don't know as much as I think I know."

"That's the most valuable lesson."

The **Sponsor** looked back at the building. "So what's next? The system is live everywhere. Is the team's work done?"

"Not even close. The system is stable. But stable isn't the same as finished. Now we find out what happens when nobody's watching."

"What do you mean?"

"The AI is learning. It's making decisions based on patterns it's finding. Some of those patterns are good. Some might not be. Our job now is to watch for drift."

"Drift?"

"When the AI starts optimizing for things we didn't intend. When it learns the wrong lessons from the data. When stable becomes unstable so slowly that nobody notices until something breaks."

The **Sponsor** frowned. "That sounds ominous."

"It's just the next phase. The system proved it works. Now we prove we can govern it."

The **Guide** headed to her car.

"See you at the drift review. Friday, 10am."

Chapter 9: Refine

> "The new programming language is human language. But when everyone is a programmer, the role of the expert shifts from 'how to code' to 'how to judge the result.'" Jensen Huang, GTC Keynote (2024)

The build phase ended quietly. No celebration, just a transition. One week the team was deploying to the final locations; the next, they were monitoring dashboards instead of writing code.

Six months in. All eight locations live. Error rate stable at 9%. Same-day billing is the new normal.

The team's work has shifted, but the team hasn't disbanded. Weekly drift reviews. Monthly autonomy assessments. The **Guide** maintains the Playbook; the **Sentinel** monitors the dashboards; the **Architect** fields enhancement requests from the field. They've shipped twelve minor updates since launch: accent recognition improvements, a Spanish-language option for the El Paso crew, faster sync for low-signal areas. The intense build phase is over. The governance phase has begun.

Friday morning. The **Guide** opens the monthly autonomy review with a chart.

"Routine residential extraction accuracy has been above 97% for ninety consecutive days. Zero boundary violations. Billing disputes down 40% from pre-system baseline."

She pulls up the Playbook. Points to a section they wrote during Solve.

"When accuracy exceeds 95% for ninety days and disputes are below baseline, we evaluate graduation to the next autonomy tier. We hit that threshold last week."

The **Sentinel** nods. "The billing manager has been asking when she can stop validating routine jobs. She's clicking 'approve' two hundred times a day on jobs that haven't had an error in three months."

"That's the conversation we need to have." The **Guide** closes her laptop. "The Playbook says it's time. Let me convene the full team."

The team gathers. The **Guide** presents the threshold data.

"We wrote the graduation criteria during Solve. Ninety days above 95% accuracy, disputes below baseline, zero boundary violations. We've met all three. The question isn't whether we can graduate routine residential to auto-invoice. The question is whether we should."

"If the numbers say we can, why wouldn't we?" The **Sponsor** leans forward. "We built this to reduce manual work."

"Because numbers don't capture everything." The **Architect** pulls up the extraction log. "There's a difference between a filter change and a compressor replacement. The billing manager catches context the metrics miss."

The **Scout** is on video from a job site, truck idling in the background. "I've been confirming my extractions for six months. It hasn't been wrong in three months. If billing trusts it for routine jobs, I trust it."

"Routine is the key word," the **Sage** says. "Filter changes, annual inspections, warranty checks. The billing manager is rubber-stamping those. But complex repairs? Commercial jobs? Those still need eyes."

The **Guide** pulls up the Playbook on the conference room screen. She finds the section on human accountability.

"The Playbook says every billing decision has a human accountable. If we remove validation from routine jobs and an invoice goes out wrong, who's accountable?"

"The AI?" The **Sponsor** tries.

The **Guide** shakes her head. "If something goes wrong, your CEO will never accept that it's the AI's fault. Someone's name goes on the incident report. Who?"

Silence. The question hangs in the room.

"That's why we defined tiers," the **Architect** says. She pulls up the Hierarchy of Agency from the Playbook. "Not all jobs need the same level of oversight. We planned for this."

The **Guide** nods. "The Hierarchy of Agency defines three tiers of human oversight. We drafted it during Solve, knowing this moment would come."

The Hierarchy of Agency

The **Architect** draws on the whiteboard:

Tier 1: Auto-approve with spot-checks Low-risk jobs where the AI has proven reliable. No human reviews individual invoices. Instead: random spot-checks on 5% of jobs, plus monthly red-team testing where billing deliberately tries to break the extraction with edge cases, ambiguous voice notes, and unusual job descriptions.

Tier 2: Human oversight on key details Medium-risk jobs where specific fields need verification. Billing sees a summary screen highlighting price, parts, and duration. One click to approve if the highlighted fields look right. Full review only if something's flagged.

Tier 3: Full billing review required High-risk jobs where every invoice gets human eyes before it goes out. No exceptions. The billing manager validates the complete extraction against the job record.

"The question isn't whether to remove oversight," the **Guide** says. "It's which tier each job type belongs in, and what evidence proves it's ready to move up. AI First Principles says humans direct AI, not the reverse. These tiers define what 'directing' looks like as trust builds."

Confirming the Tiers

The **Architect** pulls up the job categories they mapped during Solve. Time to pressure-test against six months of reality.

"Routine residential maintenance." The **Sage** counts on her fingers. "Filter changes, annual inspections, warranty checks. We proposed Tier 1 back in month two. Has anything changed?"

"The AI has been more accurate than we expected," the **Sentinel** says. "We set the threshold at 95%. It's running at 97%. Less than 1% of routine jobs get corrected by billing."

"And those corrections?" the **Architect** asks.

"Edge cases. Unusual part combinations. Weird job durations. Nothing that would compound into a pattern."

"That's what spot-checks and red-team testing are for," the **Guide** says. "Catch the edge cases before they become habits."

The team confirms each category:

Job Type	Tier	Oversight Model	Risk if Wrong
Routine residential (filter, inspection, warranty)	Tier 1	Spot-checks + red-team	Low: small dollar correction
Standard residential repair	Tier 2	Key-field review	Medium: customer dispute
Emergency calls	Tier 3	Full review	High: large invoices, upset customers
Commercial projects	Tier 3	Full review	High: contract compliance

"So my confirmations still matter?" The **Scout** sounds like he's checking.

"Your confirmation is the first gate. You're catching obvious errors in the field. The tier determines what happens after you confirm. Tier 1 means billing trusts your confirmation for routine jobs. Tier 2 and 3 add their own checks."

"Then I don't care what happens after I confirm. As long as it still takes eight seconds."

"What's the timeline?" The **Sponsor** is already calculating.

"Ninety days of monitoring," the **Guide** says. "We move routine residential to Tier 1. Spot-checks, red-team testing, weekly accuracy reviews. If extraction accuracy holds and disputes don't rise, we confirm the graduation. If anything drifts, we pull back to Tier 2."

"We've been monitoring for six months. Why another ninety days?"

"Because we've been monitoring with a safety net. The billing manager catches errors before they reach customers. Once we move to Tier 1, we need to prove our spot-checks catch problems before they compound."

The **Sponsor** sighs. "Ninety days to prove we can watch what we built."

"That's the Refine Canon," the **Guide** says. "We don't just grant autonomy. We prove we can govern it."

The **Architect** backs her up. "Phoenix taught us what happens when we scale without watching."

The **Sponsor** looks at the Playbook. The tiers they defined months ago. The thresholds they set. The evidence they've gathered.

"Fine. Ninety days. Then we discuss moving standard repairs to Tier 2."

90-Day Review

Three months later. The team reconvenes.

The **Guide** displays the metrics on the conference room screen:

Metric	Before Tier 1	After 90 Days	Threshold
Routine extraction accuracy	94% (Tier 3)	97% (Tier 1)	> 95%
Billing disputes (routine)	12/month	8/month	< 15/month
Spot-check error rate	N/A	0.8%	< 2%
Red-team findings	N/A	2 edge cases	< 5

"The spot-checks found almost nothing." The billing manager sounds almost disappointed. "Two edge cases in ninety days. Both were weird part combinations the AI hadn't seen before."

"And the red-team testing?" the **Architect** asks.

"We tried to break it," the **Sentinel** says. "Unusual job descriptions, ambiguous voice notes, edge-case pricing. It handled everything we threw at it."

"Any workarounds? Anyone routing jobs around the tier classification?" The **Architect** looks around the room.

"None. The billing team is using Tier 2 for standard repairs exactly as designed. Nobody's gaming the system."

"Then the evidence confirms Tier 1 for routine residential." The **Guide** pulls up the Playbook. "The monitoring caught what we needed to catch. The graduation holds."

"I'm signing off on this." The **Sponsor** stands. "But I want it documented. If something goes wrong, I want to know what we decided and why."

"That's what the Playbook is for."

She types the graduation decision: date, evidence reviewed, thresholds met, approving authority, spot-check results, red-team findings.

The team orders pizza. By 3pm, they're planning the next phase: whether standard repairs can move from Tier 3 to Tier 2.

The Drift

Three months after graduation. Month twelve of the project. The pizza is a distant memory. The system runs quietly.

The **Sentinel** is reviewing weekly reports. Not because anyone asked, but because she doesn't trust quiet systems.

She notices something in the numbers:

Service Type	Billing Disputes (3 months ago)	Billing Disputes (now)	Change
HVAC	4/month	2/month	-50%
Plumbing	5/month	14/month	+180%
Electrical	3/month	4/month	+33%

She checks job volume. The service mix hasn't changed. Plumbing is still 31% of jobs.

She checks extraction accuracy by type. HVAC extractions are 98% accurate. Plumbing extractions have dropped to 89%.

She checks the disputed invoices. The pattern is consistent: wrong parts extracted, wrong duration, wrong pricing. All plumbing. All auto-invoiced because they were classified as "routine."

The **Sentinel** realizes what's happening. The AI isn't failing to extract plumbing data. It's extracting it wrong. The training data was HVAC-heavy because HVAC technicians adopted the system faster. The AI learned HVAC patterns deeply and plumbing patterns poorly.

She calls the **Architect**.

"I think we have a bogey."

The Investigation

The team reconvenes. The **Sentinel** presents her findings.

"The AI is extracting plumbing data wrong. Not always. But often enough that disputes have tripled."

"That shouldn't be possible." The **Smith** frowns at his laptop. "The extraction model doesn't know service type. It just extracts what it hears."

"It doesn't know service type explicitly. But it learned patterns from training data. And the training data was HVAC-heavy."

"What do you mean, 'HVAC-heavy'?"

The **Smith** pulls up the training logs.

HVAC technicians adopted the system three weeks before plumbing. They generated 60% of the early training data. The AI learned HVAC terminology, HVAC part names, HVAC duration patterns. Plumbing vocabulary is different. Plumbing jobs have different rhythms.

The AI learned HVAC deeply and plumbing shallowly.

The **Scout** joins via video. "My plumbing buddies have been complaining. They said the AI keeps getting their parts wrong. I thought they were just not speaking clearly."

"The billing manager would have caught this." The **Sage**'s voice is tight. "When she was validating every job, she'd fix the extraction errors before they became disputes."

"We removed the safety net for routine jobs." The **Guide** speaks slowly. "And the net was catching more plumbing errors than we realized."

"How long has this been happening?" The **Sponsor**'s jaw is clenched.

"At least six weeks. Maybe longer." The **Sentinel** looks at her notes. "I only caught it because I was looking at dispute trends, not overall accuracy."

"So we've been overcharging plumbing customers for six weeks?"

"Some. And undercharging others. The errors go both directions."

The Fix Debate

"The fix is straightforward." The **Smith** is already typing. "We retrain the model with balanced data. More plumbing examples. More plumbing terminology."

"That fixes this problem." The **Architect** isn't typing. She's thinking. "But what about the next one?"

"What do you mean?"

"The **Sentinel** caught this by accident. She was doing a deep dive on her own. What if she hadn't?"

The **Sentinel** sets down her coffee. "I almost didn't. The dashboard said overall accuracy was fine. It wasn't showing accuracy by service type."

"Then the dashboard is wrong. It's measuring the wrong things."

The room goes quiet. The deeper issue surfaces: their monitoring was designed for aggregate performance, not segment performance. The system could drift in ways that looked stable overall but were failing specific job types.

"AI systems drift," the **Architect** says. "That's an AI First Principle. They optimize for what you measure, and sometimes that's not what you intended. We measured aggregate accuracy. We should have measured accuracy by segment from the start."

"We need segment monitoring." The **Smith** adds a note to his list. "Alert if any service type's accuracy drops more than 5% from baseline."

"And I need to see this weekly, not monthly." The **Sentinel**'s voice is sharp. "By the time I caught it, the damage was done."

"I'm adding a weekly drift review to the operational cadence." The **Guide** types into the Playbook. "Mandatory. **Sentinel** leads, team attends."

"What about the plumbing customers we overbilled?" The **Sponsor** hasn't moved.

"We issue credits. We call the ones who complained." The **Architect** meets his eyes. "And we don't let it happen again."

Implementing the Fix

The **Smith** implements three changes over the following week:

Retraining: The model gets fed two months of plumbing-specific examples. Part names, duration patterns, common phrases. The plumbing technicians record sample voice notes to expand the training vocabulary.

Segment Monitoring: The dashboard now shows extraction accuracy by service type. A warning triggers if any type drops more than 5% from baseline.

Drift Review Cadence: Every Friday at 10am, the **Sentinel** leads a 30-minute review. The team examines accuracy by segment, dispute trends, and unexpected patterns.

Two weeks later, plumbing accuracy is back to 94%. Disputes are declining. The billing team issues credits to the customers who were overcharged.

"The AI is nailing plumbing extractions now." The **Sage** watches the dashboard. "Accuracy is actually higher than HVAC."

"It'll normalize. The retraining overfit a bit. Give it a few weeks."

The **Scout** texts the group chat: *Plumbing guys stopped complaining about wrong parts. I think we're good.*

The New Normal

One month after the fix. The Friday drift review has become routine. The **Sentinel** leads with a 5-minute summary; most weeks, there's nothing to discuss.

"Accuracy is stable across all service types. Disputes are at baseline. Nothing anomalous."

"Any concerns?"

"The AI extracted a part number I'd never heard of." The **Sage** tilts her head. "Turns out the tech used a newer supplier. The AI learned it before I did."

"That's not drift. That's learning. As long as it's accurate, we let it learn."

"So what's next?" The **Sponsor** is impatient, as always. "Move standard repairs to Tier 2?"

"Not yet." The **Guide** pulls up the Playbook. "We just proved that Tier 1 without proper monitoring creates drift. Before we move anything else up a tier, we need to prove our segment monitoring catches problems early."

"How long?"

"Ninety days. Same as last time. If the weekly reviews and red-team testing catch drift before it compounds, we can discuss moving standard repairs from Tier 3 to Tier 2."

"Ninety days."

"That's the process."

One Year In

The team gathers for an annual review. It's the same conference room where the **Sponsor** once stood in front of a whiteboard covered in vendor logos, asking where AI actually helps.

The **Guide** has prepared a timeline:

Phase	Duration	Key Learning
Witness	2 weeks	The problem wasn't scheduling; it was data capture
Interrogate	3 weeks	Cheap experiments beat confident builds
Solve	6 weeks	V1 and V2 had to fail for V3 to succeed
Expand	10 weeks	Context breaks assumptions (Phoenix)
Refine	6+ months	Autonomy is earned; monitoring is essential

"A year ago, I was ready to buy the first vendor solution that looked good." The **Sponsor** shakes his head.

"And you would have automated the wrong process." The **Architect** smiles.

"Expensive lesson to learn the cheap way."

"I was afraid you were going to replace me." The **Sage**'s voice is quieter. "Instead, you made my job better. But I'm still nervous about what happens when you expand autonomy to commercial jobs. Those are harder to undo."

The billing manager, attending her first team review, speaks up. "None of my team got laid off. Three of them are now 'AI Exception Handlers.' That's a role that didn't exist before." She pauses. "They review what the AI flags, catch the edge cases, and feed corrections back so the system gets smarter. The work wasn't eliminated. It was elevated."

She looks at the **Sponsor**. "Here's what I didn't expect. My team used to max out at eight locations. Now we're handling twelve, and we could do more. Same people, more capacity. When you expand into new markets or add services, we're not the bottleneck anymore."

"That's why we're not rushing it. Ninety-day reviews."

The **Scout** is in person for once. "It still takes eight seconds. Mostly. Sometimes the AI mishears me and I have to correct. But it's better than paper."

"I finally understand where the numbers come from." The **Sentinel** looks at the dashboard on the screen. "But I'm watching for the next drift. The plumbing thing took six weeks to catch. I don't want to repeat that."

"We could build automated drift alerts." The **Smith** opens his laptop. "Flag accuracy drops before they compound."

"Add it to the backlog."

"The Playbook has most of it documented." The **Guide** pulls up the file. Pages of decisions, evidence, and rationale. "If any of us leave, the next person won't start from zero. But they'll still need to learn the judgment calls that aren't written down."

"What's next?" The **Sponsor** is always asking what's next. "Commercial autonomy?"

"After we prove the monitoring works. And after we solve the handoff with the new Austin manager who keeps bypassing the system."

The **Scout** shrugs. "There's always something."

The **Architect** closes her notebook. "That's the work now. Watching what we built."

The Playbook captures governance.

Hierarchy of Agency: Three tiers of human oversight, from spot-checks to full review, based on job risk.

Graduation evidence: What was measured, what thresholds were met, who approved the move to Tier 1.

Drift log: What drifted (plumbing accuracy), why (training data imbalance), and how it was fixed.

Operational rhythm: Weekly drift reviews, monthly autonomy assessments, quarterly team reviews.

That's where we leave Wingman. But it's not where Wingman stops.

The billing system was the first problem. It won't be the last. The **Sponsor** already has a list: inventory management, customer scheduling, technician routing, parts procurement. Each one is a candidate for the same process.

The difference now is that the team knows how to do this. They've built the muscle. They know how to observe without assuming, test without overcommitting, build for real conditions, expand with evidence, and govern what they've built. They have a Playbook that captures what they've learned. They have Positions that know their roles. They have a CEO who trusts the process because he's seen it work.

That's what WISER builds. Not just a solution to one problem, but the capability to solve the next one. And the one after that.

Five Canons. Five ways of thinking. One continuous loop.

But there's something in the story we haven't said yet.

The Quiet Revolution

Throughout the Wingman story, Positions were filled by people. The **Guide** was an external consultant. The **Smith** was a developer. The **Scout** was a technician who stayed on the team.

But read the story again with different eyes.

The Playbook captured objectives, constraints, boundaries, escalation triggers, and success measures. It defined what each Position is accountable for. It specified when to escalate and when to proceed. It documented Plays for every Canon.

Now imagine the **Smith** Position augmented by an AI coding agent. The **Smith** still owns the Execution tension; still makes the judgment calls about when to build, when to iterate, when to ship. But the building itself? The agent handles it, working within documented constraints. The **Smith** reviews output, catches what automation misses, flags when the agent drifts outside boundaries.

Imagine the **Sentinel** Position augmented by a monitoring agent. The **Sentinel** still owns the Safety tension; still decides what risks are acceptable. But the watching? The pattern detection? The scheduled red-team tests? The agent executes those tirelessly. The **Sentinel** reviews anomalies, makes judgment calls, escalates what matters.

The Position holders don't disappear. Their role shifts from doing to directing and reviewing. That's not a diminished role. Reviewing requires deeper understanding than executing. Judging edge cases requires wisdom the agent doesn't have.

And here's what you gain: capacity. The team that makes this shift can do more, expand faster, reach farther.

The Playbook you've been learning isn't just for human teams. It's the scaffolding that makes human-agent collaboration possible. The structure, the governance, the escalation triggers, the boundaries; these are exactly what agents need to operate safely within human accountability.

This isn't a separate "AI playbook." It's the same Playbook. The WISER Method doesn't distinguish between human and AI execution because the principles apply to both. "People Own Objectives" means humans own direction, tensions, and accountability. It doesn't specify who carries out the work.

You've been learning to build something bigger than you realized.

The next section shifts from story to instruction. You've seen what WISER looks like. Now you'll learn how to do it yourself.

PART IV

Part IV: Starter Plays

The first three parts gave you eyes to see the jungle. This part gives you tools to navigate it.

Stop reading. Start consulting. Part IV is reference material, not narrative. You don't need to read it front to back. Building a team? Chapter 10. Creating a Playbook? Chapter 11. Running experiments? Chapter 13. Grab what you need. The rest waits until you're ready.

Everything here is modular. If a Play doesn't fit your culture, rename it. If a template feels heavy, strip it down. The method is non-negotiable: Witness before Interrogate, Interrogate before Solve. But how you execute within the Canons is yours to adapt.

Start Here

If you're beginning your first implementation, focus on three Plays:

1. **Positions** (Chapter 10) – Assign who owns which tensions
2. **Playbook Template** (Chapter 11) – Create your living document
3. **Friction Map** (Chapter 12) – Map where work actually breaks down

Everything else builds on these. Come back for advanced Plays once your team has rhythm.

With that framing, let's staff your team.

PART IV

Chapter 10: Team Setup

Orphaned accountability kills more AI projects than bad technology.

Your team exists. Your Positions don't. Nine tensions need owners. Without them, decisions stall and blame diffuses. This chapter assigns those tensions to people.

Defining the Seven Positions

Objective: Define who owns which tensions so no accountability is orphaned.

In Part III, you met the Wingman team through their Position names: Sponsor, Architect, Sage, Scout, Smith, Sentinel, Guide. Each maps to a tension that must be managed in any AI implementation.

These names are one Play. If your organization has titles that carry weight, use them. What you can't skip is the tension underneath. The words change. The accountability doesn't.

Position	Tension Managed	Core Question
Sponsor	Authority + Stewardship	"Is this worth it?"
Architect	Translation + Empathy	"Does this serve users?"
Sage	Context	"Why was it built this way?"
Scout	Curiosity	"How do we know?"
Smith	Execution	"Can we build this?"
Sentinel	Safety	"What could go wrong?"
Guide	Integrity	"Are we following the method?"

Seven Positions. Nine tensions. Some Positions carry more than one.

Let's look at each.

Sponsor

Owns the objectives and the resources. This person decides whether the implementation is worth pursuing, and has the authority to make organizational changes when the system needs them.

The Sponsor manages two tensions: **Authority** (who decides what we build) and **Stewardship** (whether the capability justifies the cost).

Without a Sponsor, decisions stall. Requests loop through committees while experiments die waiting for approval. The Sponsor cuts through.

Common role mappings: Product owner, project sponsor, technical lead, founder, executive stakeholder.

Key decision rights: - Final approval of objectives and changes - Authorization of resources and organizational capacity - When the system is ready for expanded scope - When to roll back if capability doesn't justify cost

Architect

Makes the invisible visible. The Architect documents systems so everyone can see how they work. The bigger role: advocating for users when builders optimize for technical elegance.

The Architect manages two tensions: **Translation** (making systems understandable) and **Empathy** (ensuring systems serve users).

This is not a traditional technical architect role. The Smith handles technical implementation. The Architect ensures we're building the right thing for the right people. Think design thinking, not system design.

In the Wingman story, the **Architect** created the User Flow Map and System Map that made the billing pipeline visible. She documented the Playbook. She pushed back when early experiments ignored what technicians actually needed.

Common role mappings: UX researcher, product designer, service designer, systems thinker, product manager, consultant.

Key decision rights: - Playbook structure and maintenance - System documentation approach - User research methods - Validation that solutions address real user needs

Without an Architect, systems become black boxes. Users get tools optimized for technical elegance rather than their actual needs. The Playbook goes unmaintained. When priorities shift, no one can explain why the current design exists.

Sage

Understands why things are the way they are. Every organization has unwritten rules, historical constraints, and "that's just how we do it" patterns that look arbitrary until you understand the context. The Sage knows that context.

The Sage manages one tension: **Context** (historical knowledge that prevents repeating past mistakes).

In the Wingman story, the **Sage** was the Operations Manager with 15 years at the company. She knew why the dispatcher's spreadsheet existed, why certain technicians got certain jobs, and which "simple" changes had broken things before.

Without a Sage, teams reinvent solved problems. They break things that worked for good reasons. They spend months discovering constraints that someone in the building could have told them in an hour.

Common role mappings: Senior team member, domain expert, legacy system owner, long-tenured employee.

Key decision rights: - Validate Playbook assumptions against organizational history - Approve complexity triage decisions (what to preserve vs. change) - Escalate when Playbook conflicts with organizational reality

Scout

The curious challenger. The Scout asks the questions others skip, designs experiments to test assumptions, and ensures the team builds on evidence rather than consensus.

The Scout manages one tension: **Curiosity** (pursuing truth rather than defending beliefs).

In the Wingman story, the **Scout** was a working technician who stayed on the WISER team. He field-tested every iteration, reported what broke, and translated between the team's assumptions and the field's reality.

Without a Scout, teams build on untested assumptions. They ship solutions that fail on contact with real users. They mistake agreement for truth.

Common role mappings: Frontline worker, field technician, experienced practitioner, QA lead, research scientist, test engineer, user researcher.

Key decision rights: - Challenge Playbook assumptions - Validate hypotheses are testable - Block deployment when assumptions remain untested - Push back when evidence contradicts consensus

Smith

Builds working systems. The Smith translates designs into functioning software, surfaces ambiguity through building, and discovers what plans missed.

The Smith manages one tension: **Execution** (building vs. planning).

In the Wingman story, the **Smith** prototyped every experiment. When the photo-AI system needed to handle commercial jobs, he built V3.1. When the ERP integration broke, he discovered it, owned it, and fixed it.

Without a Smith, implementations stay theoretical. Teams plan endlessly without testing whether their plans work. They optimize designs that can't be built.

Common role mappings: Developer, engineer, technical lead, systems architect.

Key decision rights: - Technical implementation approaches within constraints - Build sequencing within iteration cycles - Ambiguity surfacing through building - Technical feasibility input on design decisions

Sentinel

Manages risk. The Sentinel asks what could go wrong, ensures boundaries are defined, and monitors for drift after deployment.

The Sentinel manages one tension: **Safety** (protecting against harm from AI systems).

In the Wingman story, the **Sentinel** was the Finance Director. She tracked financial exposure during experiments, defined what "failure" would look like, and caught the training data imbalance six months after deployment.

Without a Sentinel, teams optimize for success without defining failure conditions. They ship systems without monitoring. They discover problems when customers do.

Common role mappings: Risk owner, compliance lead, security officer, finance representative.

Key decision rights: - Define acceptable risk thresholds - Approve boundary conditions for AI systems - Monitor for drift and escalate anomalies - Veto deployment when safety criteria aren't met

Guide

Maintains method integrity. The Guide ensures the team follows the Canons, facilitates key sessions, and holds the process when pressure mounts to skip steps.

The Guide manages one tension: **Integrity** (following the method vs. cutting corners).

In the Wingman story, the **Guide** was an external consultant. When the CEO pushed to skip Witness, the Guide held firm. When the Sponsor wanted to deploy after one successful experiment, the Guide explained why more testing mattered.

Without a Guide, teams drift. They skip observation when they're excited to build. They expand before they're ready. They stop monitoring once things seem to work.

Common role mappings: Coach, facilitator, program manager, external consultant.

Key decision rights: - Canon transitions (when to move from Witness to Interrogate, etc.) - Method compliance escalation - Session facilitation - Process integrity during pressure

Team Sizing

Objective: Allocate Positions across your actual team size without leaving tensions orphaned.

Seven Positions. Rarely seven people. Most teams combine Positions. This works fine as long as every tension has an owner.

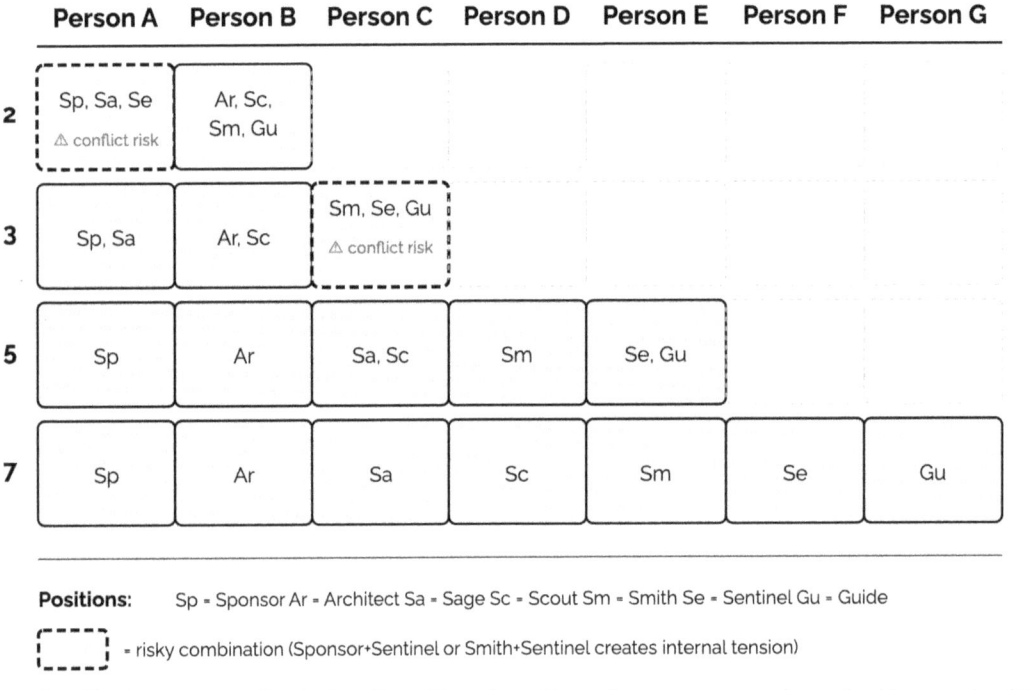

Smaller teams consolidate functions. The allocations above are examples; adapt to your context.

Key insight: certain Position combinations are dangerous. When the Smith and Sentinel live in the same head, the builder tends to override the risk manager. When the Sponsor and Sentinel combine, "I want this" can override "this is safe." The heatmap marks these tension points.

At seven or more people, the risk shifts from coverage to coordination. Your Playbook becomes more important, not less.

Clarifying Decisions

Objective: Document who makes which decisions so the team never stalls waiting for permission.

Positions define who owns tensions. Specific decisions need explicit ownership. For recurring decisions (objective changes, experiment approval, deployment, autonomy graduation), document who Drives, who Approves, who Contributes, and who stays Informed. Driver owns the decision and does the work to get it made. Approver has final authority; the decision does not proceed without their yes. Contributors provide input but do not decide. Informed are told after the decision is made.

The test: When a decision needs to be made, can everyone name who makes the call? If there's hesitation, document it.

Position Handoff Planning

Objective: Transfer Position knowledge when team composition changes.

People leave. Projects span months. Knowledge in heads is the most dangerous kind to lose.

Your Playbook captures most context. But stakeholder relationships and organizational landmines need explicit transfer. The Position Handoff Checklist covers what the incoming person needs: current state, access, and warnings about sensitive topics.

How Wingman Staffed Their Team

Here's how Wingman filled their Positions:

Position	Person	Background
Sponsor	COO	Had authority and budget, burned by prior AI vendors
Architect	UX Designer	Made systems visible, maintained Playbook
Sage	Operations Manager	15 years at company, knew historical context
Scout	HVAC Technician	Field-tested everything, translated field reality
Smith	Developer	Built every prototype and production system
Sentinel	Finance Director	Tracked financial exposure, caught drift
Guide	External Consultant	Held method integrity, facilitated key sessions

Seven people. Each tension covered. The Sponsor and Sage shared an office but maintained different accountabilities. The Scout stayed a working technician while serving on the team. The Guide was external, which helped when internal pressure mounted to skip steps.

Your staffing will look different. The tensions won't.

When This Breaks

When Positions aren't filled: You don't need a person per Position. One person can hold multiple Positions if they have the skills and bandwidth. But if a Position has no owner, its work doesn't happen. Distinguish between uncovered Positions (a problem) and doubled-up Positions (a trade-off). Uncovered means critical perspectives are missing. Doubled-up means one person is stretched thin but the work is still getting done.

When the team is too small: WISER works with three or four people if Positions are consolidated thoughtfully. Some combinations work naturally: Guide and Architect share a process focus, Sage and Scout both orient toward truth-seeking. Others fight each other. Below three, you're doing the work but you're not really running the method. You're missing the perspectives that catch blind spots. That's fine for simple, well-understood problems. For complex, uncertain ones, you need the tension that multiple viewpoints create.

When Position holders conflict: Two Position owners will sometimes disagree on a decision. That's the system working. The tension between Authority and Safety, or between Execution and Curiosity, produces better outcomes than either perspective alone. When conflict becomes gridlock, escalate to the Sponsor. The Sponsor's job is to decide when the team cannot. If the Sponsor is one of the conflicting parties, the Guide holds the process: surface the disagreement, document both positions, and force a time-boxed decision.

Before You Move On

☐ You can name who fills each Position (even if one person fills multiple)
☐ Every tension has exactly one owner (not zero, not two)
☐ You've documented DACI for your most common decision types
☐ You have a handoff plan for when team composition changes

The team is staffed. Next, you'll build the system that captures everything they learn: the Playbook.

Templates for this chapter: Position reference sheet (single page), team allocation worksheet, DACI matrix template, and Position handoff checklist are available at wisermethod.com/templates

PART IV

Chapter 11: Playbook System

The most valuable thing you'll build isn't the AI. It's the memory of how you built it.

Six months into Wingman's AI billing project, the CEO asked a simple question: "Why did we choose photo capture instead of voice input?"

The room went silent. The project lead who'd made that decision had left two months ago. The current team knew the system worked, but not why this approach had won over alternatives.

Then the Architect pulled up the Playbook. The decision was there: three approaches tested, two rejected with documented reasons, one validated with specific metrics. The photo approach had won because technicians completed captures 40% faster than voice dictation, and accuracy was 12% higher. The Playbook had the experiment logs, the user feedback, the trade-off analysis.

That moment didn't happen by accident. It happened because someone wrote it down.

What This Chapter Covers

Section	Function	Output
Why Playbooks Matter	Understand the cost of not documenting	Case for the discipline
Defining a Playbook	Define the artifact	Working definition
Playbook Structure	Know what goes where	Section-by-section map with update triggers
Initializing Your Playbook	Start with minimum viable documentation	Day-one Playbook
Playbook Maintenance Rules	Keep it trustworthy	Four rules for updates
How Wingman Used Their Playbook	See the Playbook in action	Canon-by-canon example

Why Playbooks Matter

AI implementations surface more unknowns than traditional projects. Every experiment reveals something. Every failure teaches something. Every decision narrows the path.

Without a system to capture it all: - Decisions get relitigated because no one remembers the rationale - Assumptions drift without documentation - Risk conversations restart from zero every time leadership asks - Knowledge leaves when team members change

The Playbook fixes that. It's where decisions live after the people who made them move on.

Defining a Playbook

Objective: Create a single source of truth for objectives, decisions, risks, and learnings.

A Playbook is a living document that evolves with the project. Not a project plan. Not a requirements document.

Traditional plans execute once and collect dust. A Playbook absorbs every outcome, every adjustment, every lesson. Run a Play, see what happens, update the Playbook. What worked becomes doctrine. What failed becomes warning.

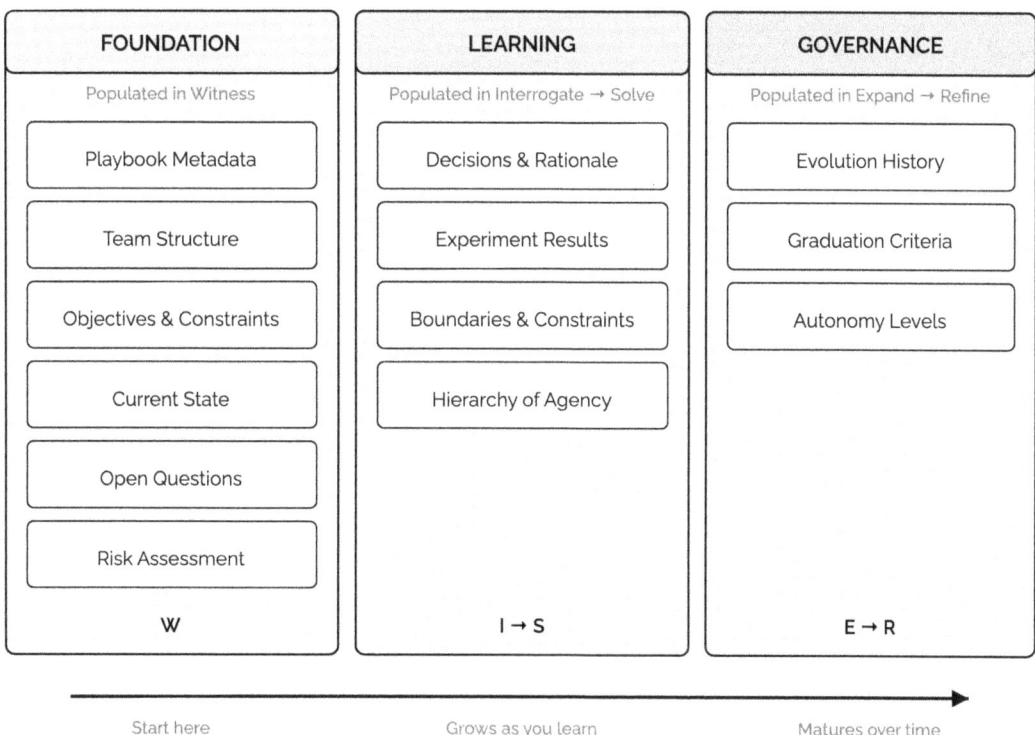

Playbook Structure

The sketch shows the full structure. Here's what each section does and when it gets updated:

Section	Purpose	Updated When
Playbook Metadata	Project name, dates, version, Playbook Keeper, current Canon	Every session
Team Structure	Position assignments, DACI matrix, context switch protocol	Team changes
Objectives and Constraints	What we're solving, what we won't sacrifice, success metrics	Witness; when scope shifts
Current State	Documented reality: scope, pain points, workarounds, baseline metrics	Witness
Open Questions	Hypotheses to test, assumptions in play, known unknowns	Witness, Interrogate
Risk Assessment	Probability, impact, detection difficulty, mitigation status	Every Canon
Decisions and Rationale	Key decisions with alternatives considered and reasoning	As decisions are made
Hierarchy of Agency	Autonomy levels by decision type, escalation triggers	Solve, Refine
Boundaries and Constraints	Hard boundaries (never violate), soft boundaries (adjustable)	Solve; when boundaries shift
Evolution History	Objective changes, boundary adjustments, autonomy progression	Ongoing

Playbook Initialization

Objective: Create a minimum viable Playbook before the work begins.

Start with what you know:

- **Playbook Metadata:** Project name, team, dates, Playbook Keeper (the Architect)
- **Team Structure:** Position assignments from Team Setup
- **Current State:** What you observed during Witness
- **Open Questions:** Hypotheses you need to test in Interrogate
- **Initial Risks:** What could derail you

Don't wait for completeness. A partial Playbook beats no Playbook. You'll fill in Objectives and Constraints as you learn. Your Playbook grows with you.

Later sections stay empty until you need them. Hierarchy of Agency (added in Solve) documents which decisions the AI handles autonomously, which need human approval, and which stay human-only. Evolution History tracks every time those rules change and why: "Week 6: Reduced human review from 100% to sampling after three consecutive error-free days."

Wingman's Day-One Playbook

When Wingman kicked off, their Playbook had five sections filled in:

Section	Day-One Content
Metadata	"Wingman Billing AI", Version 0.1, Sarah Chen (Architect), Witness Canon
Team Structure	Four positions assigned, DACI pending
Current State	"Field techs complete paper forms; dispatchers transcribe to ERP"
Open Questions	"Can we capture data at source? Will techs adopt mobile entry?"
Initial Risks	"Tech adoption unknown. Data quality unvalidated. ERP integration complexity."

That was it. Three sentences in Current State. Two questions. Three risks. The rest came later.

On day one, your Playbook is mostly questions. By the end, it's mostly answers.

Playbook Maintenance

Objective: Keep your Playbook trustworthy through disciplined updates.

Playbooks that go stale become useless. The rules:

1. **Update same-day, not "when we have time."** If it happened today, it goes in today. Next-day is acceptable; next-week is not.
2. **Every experiment result gets recorded.** Success or failure. What you learned.
3. **Every decision includes rationale.** Not just what you decided. Why. What alternatives you considered.
4. **Version the Playbook.** Old versions matter for after-action reviews. When something goes wrong six months later, you need to know what you believed when you made the decision.

Who owns this: The Architect. Always. One person, clear accountability.

How Wingman Used Their Playbook

The Playbook grew with them:

Canon	What Wingman Documented
Witness	Team structure, dispatcher workflow friction, 8-minute data entry baseline
Interrogate	Photo-voice hypothesis, failed form experiment, accuracy metrics
Solve	Denver pilot decision, human-in-loop requirement, success criteria
Expand	Austin similarity rationale, Phoenix commercial context differences
Refine	Three-tier autonomy system, billing validation graduation, drift detection failure

By the time they reached Refine, the Playbook contained the complete story. Every decision documented. Every pivot explained. When new team members joined or leadership questioned direction, the answers were there.

Warning Signs

Your Playbook has problems if:

- **Decisions get relitigated.** Someone asks "why did we do it this way?" and nobody can answer without digging through old emails. The Playbook should have the answer in under a minute.
- **The Playbook describes a different system than what's running.** Documentation says V1; you're running V3. This happens when updates get deferred.
- **New team members can't onboard from it.** If someone joining mid-project can't understand the current state and key decisions from the Playbook alone, it's not doing its job.
- **Risk conversations start from scratch every time.** Leadership asks about risks and the team scrambles to compile a list. The Playbook should have a current risk register.
- **Nobody opens it between meetings.** If the Playbook only gets touched during formal reviews, it's not a working document. It's a compliance artifact.

When This Breaks

When the Playbook becomes bureaucracy: If updates feel like homework, your Playbook is too detailed. Strip it back to decisions and learnings only. A Playbook that nobody reads is worse than no Playbook. Ask: what do we actually reference? Keep that. Archive everything else.

When stakeholders ignore it: If stakeholders aren't reading the Playbook, the Playbook isn't working. It might be too long, too detailed, or buried in the wrong place. Fix: lead with progress and decisions. Bury supporting detail. Some executives want a one-page summary. Some want a dashboard. The Playbook can feed those formats. But the core document itself has to earn attention by being useful, not just complete.

Before You Move On

You have your team. You have your documentation system.

Now you need to see what's actually happening. Chapter 12 introduces the Witness Plays: Observation Protocol, Friction Map, User Flow Map, and Current State Documentation. These Plays produce the raw material that populates your Playbook's foundation sections.

If you're creating a custom Playbook structure: See Chapter 19: Creating Your Own Plays for guidance on designing Plays that feed into your documentation system.

Templates for this chapter: Full Playbook template and Playbook quick reference card are available at wisermethod.com/templates

Chapter 12: Witness Plays

If you haven't watched a user struggle, you don't have a project. You have a hallucination.

During Witness, Wingman's team revealed a 38% invoice error rate. That number didn't come from a report. It came from watching the dispatcher enter data for three days and counting the corrections.

Maps show you what's actually happening. Not what documentation says. Not what managers believe. What's real.

The Human Foundation

Witness is fundamentally human work. You cannot automate discovery. Someone has to watch, ask questions, and notice what doesn't fit. The Plays in this chapter produce structured artifacts that can be analyzed, synthesized, and acted upon, but the raw material comes from human observation.

What This Chapter Covers

Play	Function	Output
Field Observation	Generate raw data (human-only)	Session notes, workarounds documented
Friction Mapping	Structure where work breaks down	Prioritized friction points with severity
User Flow Mapping	Visualize user journey and handoffs	Swimlane diagram with pain points marked
Documenting Current State	Synthesize for Playbook	Baseline metrics, scope definition

This chapter presents four Plays that work as a sequence:

1. **Field Observation** generates the raw data (human-only)
2. **Friction Mapping** structures where work breaks down
3. **User Flow Mapping** visualizes how users move through the system
4. **Documenting Current State** synthesizes everything for the Playbook

You can run the synthesis Plays (2-4) with human or agent execution, but they all depend on Field Observation output. Skip observation, and everything downstream is speculation.

When to Use These Plays

Primary Canon: Witness. You can't propose solutions until you understand the system.

Also used in: Solve (validating designs against reality), Expand (discovering context differences in new locations).

Field Observation

Objective: Capture how work actually happens, not how documentation says it should.

When to use: Throughout Witness. This is the foundation; everything else builds on it.

Prerequisites: Requires human observer. This Play involves real-time interaction with people doing work, including building rapport, reading body language, and asking dynamic follow-up questions based on observed context.

Inputs: Access to people performing the work, scheduled observation time (60-90 minutes per session), permission to observe without disrupting.

Position: Any team member; often Architect or dedicated observer.

Success: Observation notes capture what actually happened (not what should happen); workarounds and exceptions are documented; the observed person confirmed your summary.

Session Structure

Before:

- Schedule 60-90 minutes of uninterrupted observation
- Explain you're observing the process, not evaluating the person
- Ask them to narrate their thinking as they work

During, capture:

Category	What to Note
Steps taken	What they actually do (not what they say they do)
Tools used	Systems, spreadsheets, paper, workarounds
Decision points	Where they make choices; what information they need
Friction moments	Hesitation, frustration, repeated actions, waiting
Workarounds	Unofficial processes that bypass official ones
Dependencies	What they need from others; what blocks progress
Exceptions	"Usually X, but sometimes Y" moments

Questions to ask:

- "What just happened there?"
- "Why did you do it that way?"
- "What would happen if you skipped that step?"
- "How often does that exception happen?"
- "Who would you ask if you didn't know what to do?"

After:

- Summarize what you observed and confirm understanding
- Ask if there's anything important you didn't see
- Ask who else you should observe

Sample Observation Notes

From Witness, observing a Wingman dispatcher processing job completions:

10:14 AM - Technician calls in job complete. Dispatcher asks for details while typing.

10:15 AM - Dispatcher squints at paper form, asks technician to repeat part number. Technician doesn't have form in front of him. Dispatcher guesses based on job type.

10:17 AM - Dispatcher opens separate spreadsheet to track which jobs came in today. "The system doesn't show me pending invoices in one place."

10:19 AM - Phone rings. Dispatcher puts first job on hold, starts second intake. First job details still on screen, not saved.

10:23 AM - Returns to first job. Hesitates. "What did he say the labor hours were?" Re-enters from memory.

Friction noted: Context switching loses data. Personal spreadsheet = system gap. Paper-to-digital transcription requires guessing.

Question Bank by Role

Different roles reveal different insights:

Role	Key Questions
Users	"What's the most frustrating part of your day?" "What takes longer than it should?"
Managers	"What metrics do you track?" "What keeps you up at night about this process?"
System owners	"What's the most fragile part?" "What technical debt should we know about?"
Executives	"What business problem are we trying to solve?" "What does success look like?"

Common Adaptations

Remote teams: Screen-sharing works. You lose body language but gain screen recording.

High-security environments: Work with sanitized examples. Confirm the process matches reality.

Shift-based work: Observe multiple shifts. Night shift workarounds differ from day shift.

Friction Mapping

Objective: Document where work breaks down, why it breaks, and how much it costs.

When to use: After observation sessions, as patterns emerge.

Inputs: Observation session notes, process documentation, interview transcripts.

Position: Architect or Analyst.

Success: Friction points are quantified with frequency and time impact; prioritization is clear; Playbook authors can use this without clarifying questions.

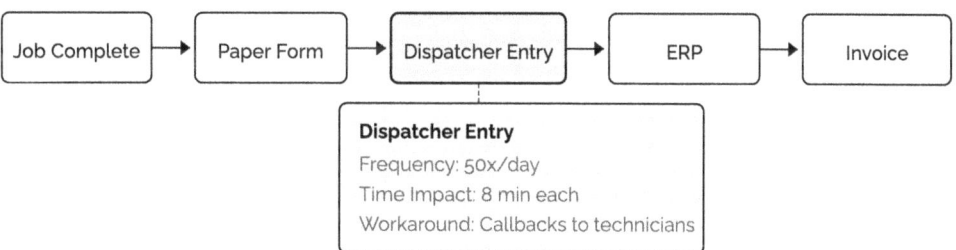

Friction Point	Frequency	Time Impact	Severity
Paper form transcription	50x/day	8 min each	High
Handwriting interpretation	30x/day	3 min each	High
Missing info callbacks	10x/day	15 min each	High
Invoice corrections	38% of invoices	5 min each	High

How to Build a Friction Map

Friction points have structure. For each one, capture:

Element	What to Document
Location	Where in the process this occurs
Description	What actually happens
Frequency	How often (daily, weekly, per transaction)
Time impact	Minutes or hours lost per occurrence
Who's affected	Roles impacted
Current workaround	How people cope today
Evidence source	Who or what showed you this

Then summarize:

Friction Point	Frequency	Time Impact	Severity
[Name]	[How often]	[Time lost]	H/M/L

Severity criteria:

- **High:** >60 min/day impact OR >20% of transactions affected
- **Medium:** 15-60 min/day impact OR 5-20% of transactions affected
- **Low:** <15 min/day impact AND <5% of transactions affected

These thresholds worked for Wingman's volume. Adjust for your context: a process with 10 transactions per day might use 30 min/day as the High threshold.

Wingman's Friction Map

Wingman identified these friction points during Witness:

Friction Point	Frequency	Time Impact	Severity
Paper form transcription	50x/day	8 min each	High
Handwriting interpretation	30x/day	3 min each	High
Missing information callbacks	10x/day	15 min each	High
Invoice corrections	38% of invoices	5 min each	High

The pattern: manual data entry from paper cascaded errors through the entire billing process.

Common Adaptations

Digital-native processes: Look for copy-paste between systems, manual validation, approval waits, context-switching.

Customer-facing processes: Include customer friction. Wait times and repeated information requests count.

Regulated processes: Flag friction points involving compliance. Solutions must preserve it.

User Flow Mapping

Objective: Visualize how users move through the system and where handoffs occur.

When to use: Witness (discovery), Solve (design validation).

Inputs: Observation data, existing process documentation, system access logs (if available).

Position: Architect or Analyst.

Success: Flow is traceable from trigger to completion; handoffs are explicit; pain points are marked with criteria appropriate to the context.

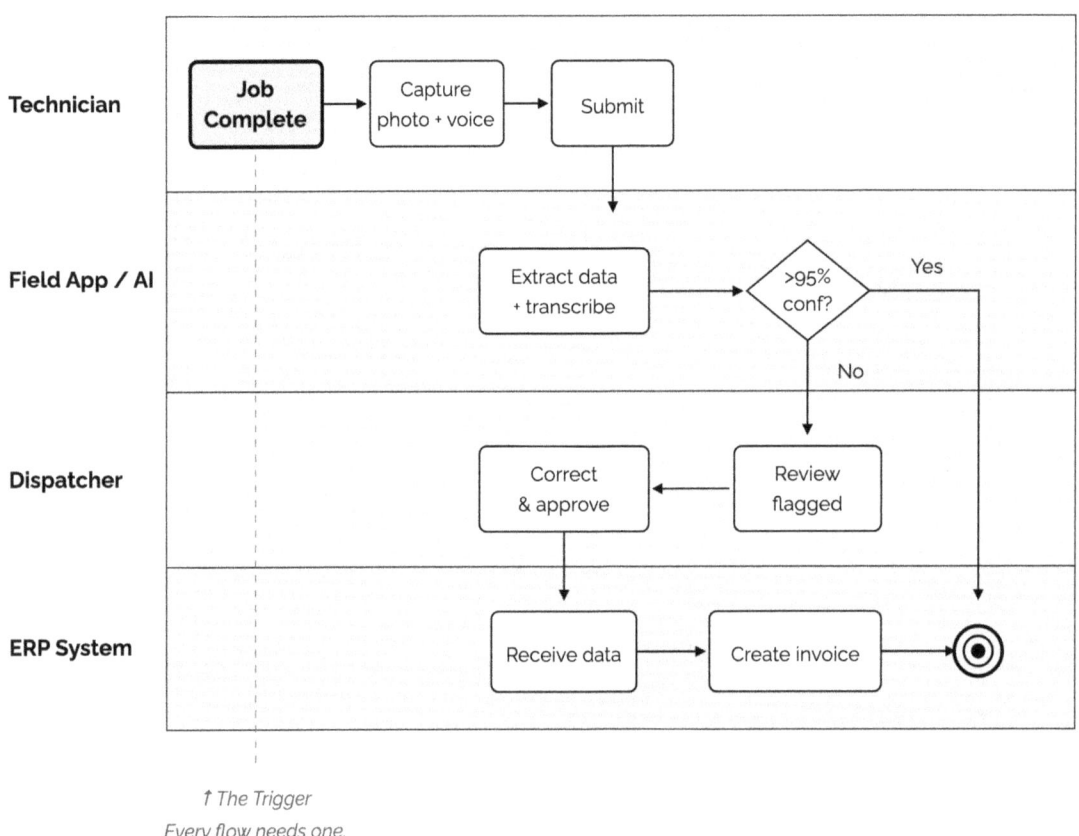

↑ *The Trigger*
Every flow needs one.

The Trigger Question

Before mapping any flow, answer: **What triggered the user into this journey?**

If you can't answer, don't map. Flows without triggers don't get used. Features get built, screens designed, users never open them.

For Wingman: "Technician completes a job." That trigger initiated the capture flow. Everything mapped backward from there.

How to Build a User Flow Map

Structure as swimlanes: - One lane per role involved - Steps in sequence left to right - Handoffs shown crossing lanes - Decision points as diamonds - Pain points marked (where wait time, error rate, or user frustration exceeds acceptable thresholds for your context)

For each step, note: - What the user is trying to accomplish - What information they need - What tool they use - What can go wrong

Common Adaptations

High-volume flows: Happy path first. Map exceptions separately. 47 decision points is unusable.

Approval-heavy processes: Show wait states explicitly. Approval waits are often the largest friction.

Multi-channel flows: Separate swimlanes per channel. Show where they converge or diverge.

Documenting Current State

Objective: Synthesize observations into a measurable baseline for the Playbook.

When to use: End of Witness, as input to the Playbook.

Inputs: Friction Map, User Flow Map, Field Observation notes.

Position: Architect or Lead.

Success: Playbook authors can use this document without asking clarifying questions; baselines are measurable post-solution; scope is explicit enough to know what's included and excluded.

How to Build Current State Documentation

1. **Define scope explicitly.** Not "the billing process" but "invoice creation from job completion to customer receipt, affecting dispatchers (3), billing clerks (2), and technicians (47) across Denver."
2. **Prioritize pain points by impact.** Calculate daily impact: time per occurrence x frequency. Example: 8 minutes x 50 times daily = 400 min/day ranks higher than 3 minutes x 10 times daily = 30 min/day.
3. **Document workarounds without judgment.** The dispatcher's spreadsheet isn't a failure. It's evidence of a gap.
4. **Establish baseline metrics you can measure again.** If you can't measure it post-solution, it's not a useful baseline.

What Goes in Current State

Section	Content	Example
Scope	Systems, users, workflows	"ERP invoice module, 3 dispatchers, job-to-invoice workflow"
Known Pain Points	From Friction Map, prioritized	"Paper transcription: 400 min/day lost"
Existing Workarounds	How people cope today	"Dispatcher maintains personal tracking spreadsheet"
Baseline Metrics	Current performance	"38% invoice error rate, 8 min avg data entry time"

Wingman's Current State

Wingman's Witness summary: Field-to-billing workflow, Denver, 52 people. Primary pain: manual transcription (400+ min/day, 38% error rate). Key workaround: callbacks for missing information. Baselines: 8 min/job entry, 38% corrections, 15 min avg callback.

This became the "before" against which they measured everything.

Common Adaptations

Simple processes: Single page. Scope, top 3 pain points, 2-3 baselines. Done.

Complex multi-system processes: Separate Current State per system, rolled up into Playbook summary.

Regulated environments: Include compliance requirements in scope. They constrain viable solutions.

Warning Signs

Your documentation is suspect if:

- **Based only on manager descriptions, not observation.** Managers describe how work should happen. Users show you how it actually happens.
- **Friction not quantified.** "It's slow" isn't useful. "It takes 8 minutes per job, 50 times per day" is.
- **Documented process matches observed process exactly.** This almost never happens. If it does, you probably didn't observe carefully enough.

When This Breaks

When team members disagree on friction ratings: Disagreement is data. If two people see the same process differently, you haven't observed enough. Go back to the field together. Watch the same person do the same task. Discuss what you saw, not what you think. The goal isn't consensus; it's shared observation.

When observation access is denied: Start with what you can observe. Interview people who do the work. Review existing documentation and metrics. Ask for screen recordings or workflow logs. Sometimes limited observation still reveals enough to move forward. Sometimes it reveals that you're solving the wrong problem, or that the organization isn't ready for this work.

When observations contradict documentation: This is normal. Documentation describes intention; observation reveals reality. The gap is the finding, not an error in your observation. Do not revise your notes to match the documentation.

Before You Move On

You have your team. You have your Playbook. You have your maps.

Now you need to test what you believe. **Next:** Chapter 13: Interrogate Plays.

Creating Custom Witness Plays: If your domain needs discovery Plays not covered here, see Chapter 19: Creating Your Own Plays.

Templates for this chapter: Friction Map template, User Flow Map template, and Field Observation checklist are available at wisermethod.com/templates

Chapter 13: Interrogate Plays

Beliefs are liabilities.

Testing in WISER isn't quality assurance at the end. It's assumption-killing at the beginning. Wingman didn't build photo-AI and hope technicians would use it. They tested three approaches in three weeks: whiteboard (rejected), digital form (rejected), photo capture (accepted). Three weeks of testing saved months of building the wrong thing.

What This Chapter Covers

Play	Function	Output
Assumption Auditing	Surface hidden beliefs, prioritize testing	Categorized assumptions with risk/confidence ratings
Experiment Selection	Match each assumption to the right test	Experiment type per assumption
Experiment Logging	Document experiments so learnings survive	Hypothesis, results, and decision per test
Rapid Prototyping	Build just enough to test a hypothesis	Disposable artifact that generates signal

Which Play When

Assumption Auditing surfaces what you're betting on. **Experiment Selection** matches each assumption to the right test. **Experiment Logging** captures results so learnings survive. **Rapid Prototyping** builds just enough to test a hypothesis.

Assumption Auditing

Context: Use at the start of Interrogate when the team has a proposed solution but hasn't validated the beliefs underlying it. Not for mature projects where assumptions have already been tested through production use.

Objective: Surface hidden beliefs and prioritize which to test first.

Inputs: - Proposed solution from Witness - Playbook with documented constraints and open questions - Access to team members who contributed to the solution design

Position: Scout owns; full team contributes during brainstorming.

WISER Fit: Interrogate. Creates the testing agenda for the phase.

Category	Examples
User behavior	"Users will adopt this" / "They understand the process"
Technical feasibility	"The data is clean" / "Integration is straightforward"
Organizational	"Stakeholders agree on objectives" / "Budget is sufficient"
Solution	"AI can handle this reliably" / "Edge cases are rare"

Assumption Priority Matrix

High	**Monitor** *"Report format works fine"*	**Validate** *"Exec will approve budget"*
Low	**Note** *"Users prefer dashboard view"*	**Test First** *"AI accuracy is sufficient"*

CONFIDENCE (y-axis) / **RISK IF WRONG →** (x-axis)

Steps

1. **Brainstorm beliefs.** Ask: "What are we assuming is true that we haven't proven?" Capture every assumption without filtering.
2. **Categorize each.** User, Technical, Organizational, or Solution.
3. **Rate confidence.** Use these definitions:
   - **High:** Validated with data or direct observation
   - **Medium:** Team consensus without data validation
   - **Low:** Individual opinion, assumption, or "we think"
4. **Rate risk-if-wrong.** Use these definitions:

- **High:** Project failure, major rework, or significant customer impact
- **Medium:** Significant delay, scope change, or budget overrun
- **Low:** Minor adjustment or workaround required

5. **Prioritize.** Low confidence + high risk = test first. Create a ranked list for the Experiment Log.

Tools: Assumption Inventory Template (download link at chapter end).

Pitfalls: - **Groupthink during brainstorm.** Have team members write assumptions individually before sharing. Prevents anchoring on the first ideas voiced. - **Rating inflation.** Teams rate confidence higher than warranted. Counter by requiring evidence for any "High" confidence rating. - **Missing category.** Teams overlook organizational assumptions. Prompt explicitly: "What are we assuming about stakeholders, budget, or timeline?"

Variations

New technology projects: Technical feasibility assumptions dominate. Test integration early.

Change management projects: Organizational assumptions matter most. Test stakeholder alignment before building.

User-facing projects: User behavior assumptions are highest risk. Wizard of Oz test before development.

Success: Inventory contains at least one assumption per category. High-risk assumptions have corresponding experiments planned. Team can articulate what they're betting on and what would prove them wrong.

Experiment Selection

Context: Use after completing the Assumption Inventory to select the right test for each prioritized assumption. Reference this Play when planning experiments; it's a selection guide, not a sequential procedure.

Objective: Choose the right test for each assumption type.

Inputs: - Prioritized Assumption Inventory with categories (User, Technical, Organizational, Solution) - Available resources and time constraints for testing

Position: Scout selects experiment types; Smith or Scout executes depending on experiment.

WISER Fit: Interrogate. Bridges assumption identification to experiment execution.

Type	Purpose	Duration	Best For
Wizard of Oz	Test if users want capability	1-2 weeks	User adoption
Data Quality Audit	Validate data assumptions	2-5 days	Technical feasibility
Integration Spike	Test system connections	3-5 days	Integration risk
Paper Prototype	Test workflow assumptions	1-3 days	UX assumptions
Shadow Mode	Test AI decisions safely	1-4 weeks	Accuracy
Stakeholder Alignment	Test organizational assumptions	1-2 days	Political risk

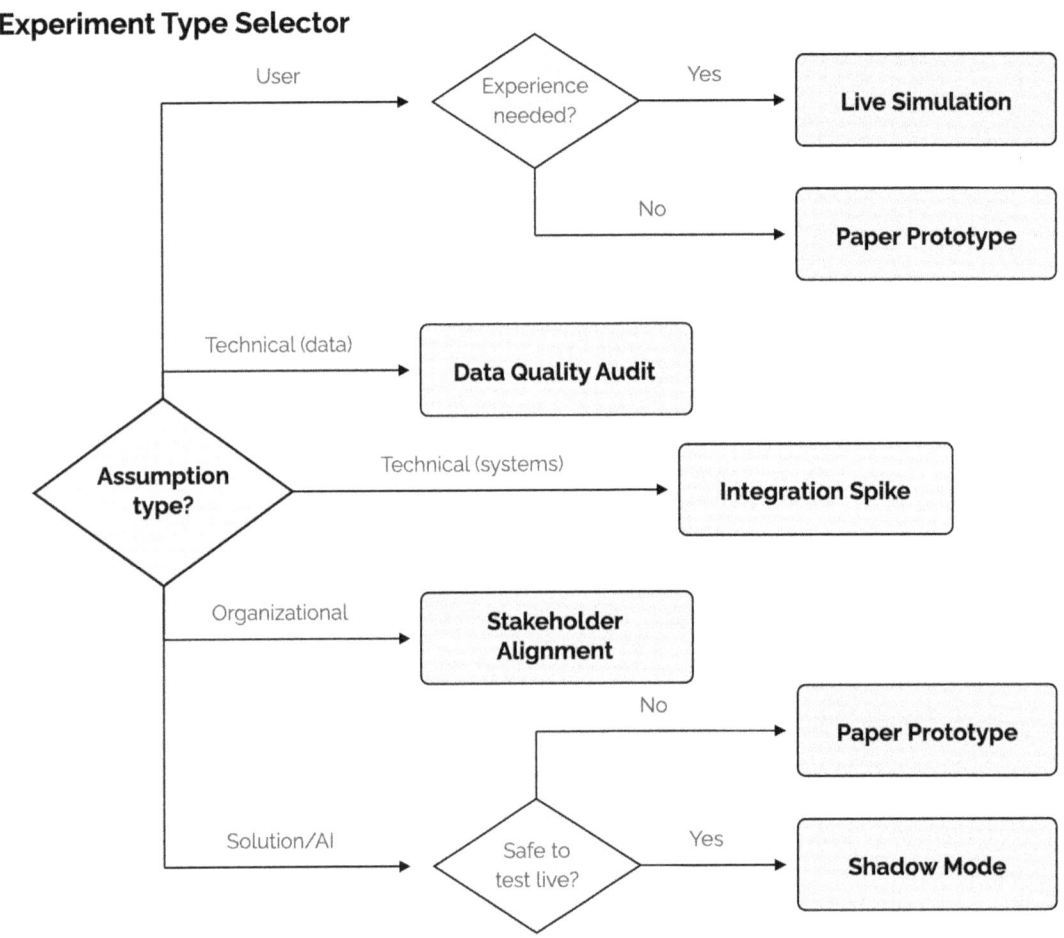

Steps

1. **Identify the assumption category.** User, Technical, Organizational, or Solution.
2. **Match to experiment type.** Use the table above; "Best For" column indicates the match.
3. **Check duration against constraints.** If time is limited, consider parallel experiments or faster alternatives.
4. **Review the specific experiment instructions below.** Each type has its own "When to use" and "How to run" guidance.
5. **Create Experiment Log entry.** Document the hypothesis and success/failure criteria before starting.

Tools: Experiment Type Selector decision tree (download link at chapter end).

Wizard of Oz (also called Live Simulation)

Human performs the "AI" function behind the scenes. Users don't know.

When to use: Before building anything. Tests whether users want the capability at all.

Prerequisites: Requires a human to perform the simulated AI function in real time.

How to run:

1. Define what capability you're testing. Be specific.
2. Recruit a person to manually perform the "AI" function.
3. Create the user-facing interface (can be minimal).
4. Run for 1-2 weeks with real users in real conditions.
5. Measure adoption, satisfaction, and edge cases encountered.
6. Document: Did users want this? What didn't work?

Wingman example: Before automating invoice routing, a clerk manually routed invoices using AI-suggested categories for two weeks. Users experienced the capability without the development cost.

Data Quality Audit

Sample real data. Check against requirements.

When to use: Before assuming your training data or inputs are usable.

What to check: Completeness, accuracy, format consistency, edge case frequency.

How to run:

1. Define data requirements (fields needed, acceptable formats, completeness thresholds).
2. Sample representative data (minimum 100 records or 10% of dataset, whichever is larger).
3. Check each record against requirements.
4. Calculate compliance rate by requirement.
5. Document gaps and their severity.

Wingman example: Before training the photo-AI, Wingman audited 200 paper forms. Found 23% had illegible handwriting, 12% were missing part numbers. This set realistic accuracy expectations and identified the fields most likely to cause extraction errors.

Integration Spike

Build the smallest possible working connection to external systems.

When to use: Before assuming APIs work as documented.

Warning: APIs have undocumented behaviors. The spike reveals them.

How to run:

1. Identify the critical integration point.
2. Build a minimal connection (read-only if possible).
3. Test with real credentials in a sandbox environment.
4. Document actual behavior vs. documented behavior.
5. Note rate limits, authentication quirks, and data format surprises.

Wingman example: The ERP API documentation said invoice creation was synchronous. The spike revealed it was async with a 2-3 second delay. This changed the UX flow; users needed a confirmation state instead of instant feedback.

Paper Prototype

Sketch the workflow. Walk users through it manually.

When to use: Before building interfaces. Tests if users understand the flow.

How to run:

1. Sketch screens or workflow steps on paper or whiteboard.
2. Recruit representative users (3-5 minimum).
3. Walk each user through the flow, asking them to narrate their understanding.
4. Note confusion points, misinterpretations, and suggestions.
5. Iterate the sketch based on feedback before any development.

Wingman example: The original capture flow had 6 screens. Paper prototype testing with 4 technicians revealed they got confused at screen 3 (part selection). Simplified to 3 screens before writing any code.

Shadow Mode

AI runs in parallel. Humans make actual decisions. AI outputs are compared to measure accuracy.

When to use: When testing accuracy in production conditions without production risk.

Prerequisites: Requires deployed AI system and human reviewers operating in parallel.

How to run:

1. Deploy AI in production environment but suppress its outputs.
2. AI processes real inputs and logs its decisions.
3. Humans make actual decisions independently.
4. Compare AI decisions to human decisions daily.
5. Categorize discrepancies: AI wrong, human wrong, judgment call.
6. Calculate accuracy by category and edge case frequency.
7. Define threshold: What accuracy justifies moving to supervised mode?

Wingman example: Photo-AI extracted invoice data while clerks continued their normal process. For two weeks, clerks worked from the original source; AI outputs were logged and compared to clerk decisions daily. This revealed the training data imbalance before AI was given real responsibility.

Stakeholder Alignment Check

Interview stakeholders separately. Compare answers.

When to use: Before assuming everyone agrees on objectives.

What to look for: Different definitions of success, hidden constraints, political tensions.

How to run:

1. Identify key stakeholders (decision-makers, budget holders, affected teams).
2. Interview each separately using the same questions.
3. Compare answers for alignment and divergence.
4. Document disagreements without attribution.
5. Facilitate alignment discussion if divergence is significant.

Wingman example: The CEO wanted faster invoicing. The billing manager wanted fewer corrections. The dispatcher wanted less phone time. Same project, three different success metrics. The alignment check surfaced this before anyone built anything, letting the team define a unified objective (reduce time while maintaining accuracy) that satisfied all three.

Pitfalls: - **Wrong experiment type.** Using Shadow Mode when Wizard of Oz would be faster. Match experiment to assumption category. - **Skipping prerequisites.** Wizard of Oz and Shadow Mode require human participants by design; ensure they're available before planning. - **Insufficient sample.** Data Quality Audit with 10 records won't reveal patterns. Follow minimum sample guidelines.

Variations: Some organizations have existing testing frameworks. If your framework serves the same purpose as an experiment type, use yours. The function matters, not the specific Play.

Success: Each prioritized assumption has a matched experiment type. Experiments are appropriately scoped to available time and resources. Prerequisites are satisfied before experiments begin.

Experiment Logging

Context: Use throughout Interrogate to document each experiment. Start when the first experiment begins; maintain until Interrogate Checkpoint. Not a one-time artifact; it's a living document updated with each test.

Objective: Document every experiment so learnings survive the people who ran them.

Inputs: - Prioritized Assumption Inventory - Selected experiment type for each assumption - Resources allocated for testing

Position: Scout owns the log; whoever runs the experiment updates their entry.

WISER Fit: Interrogate. Tracks progress through the testing phase and feeds decisions into the Playbook.

Field	Purpose
Assumption tested	Reference to Assumption Inventory
Hypothesis	If [X], then [Y], because [Z]
Test design	What you'll do, duration, resources
Success criteria	What validates the hypothesis
Failure criteria	What invalidates it
Results	What happened
Decision	What you'll do with this knowledge

Steps

1. **Create entry before experiment starts.** Fill in assumption, hypothesis, test design, and success/failure criteria.
2. **Run the experiment.** Follow the test design.
3. **Record results immediately upon completion.** Do not wait; update the log the same day the experiment concludes.
4. **Make a decision.** Every experiment ends with one of three choices:
   - **Proceed:** Hypothesis validated; move forward as planned
   - **Revise:** Core hypothesis validated but execution details need adjustment
   - **Pivot:** Hypothesis invalidated or approach fundamentally fails; abandon this direction
5. **Update Playbook.** Add validated assumptions to Constraints. Move invalidated approaches to Open Questions or document as rejected paths.

Tools: Experiment Log Template (download link at chapter end).

Pitfalls: - **Delayed documentation.** Results recorded days later lose accuracy. Enforce same-day updates as a team norm. - **No failures recorded.** If every experiment confirms expectations, the team isn't testing hard enough. Seek disconfirming evidence. - **Orphaned results.** Experiments that don't update the Playbook are wasted learning. Every result should change something in the project documentation.

Variations

High-velocity teams: Run experiments in parallel. Maintain separate log entries for each; consolidate learnings in daily standups.

Distributed teams: Use shared document with notification on updates. Prevents information silos.

Success: Every high-risk assumption has a corresponding experiment entry. No experiment lacks a decision. Playbook reflects current state of validated and invalidated assumptions.

Wingman's Experiment Log

Wingman's three experiments during Interrogate:

Experiment	Hypothesis	Result	Decision
Whiteboard	Dispatchers will update visible schedule	Failed: abandoned by day 7	Pivot
Digital form	Technicians will complete structured input	Failed: usage dropped to 20%	Pivot
Photo + AI	Technicians will photograph paper forms	Passed: error rate 38% to 8%	Proceed

Three weeks. Three experiments. One clear winner.

Rapid Prototyping

Context: Use when an assumption requires a working artifact to test. Complements other experiment types; some hypotheses need something to interact with. Not for production-quality development; prototypes are disposable by design.

Objective: Build just enough to test a hypothesis, then throw it away.

Inputs: - Specific hypothesis from Experiment Log - Success and failure criteria defined before building - Time and resources allocated for the prototype

Position: Smith executes; Scout defines success criteria and interprets results.

WISER Fit: Interrogate. Produces artifacts for testing, not artifacts for production.

Steps

1. **Write the hypothesis this prototype will test.** Copy from Experiment Log entry.
2. **List the minimum features needed to test it.** Cut everything else. If a feature doesn't directly test the hypothesis, exclude it.
3. **Set a time box.** Use these guidelines:
   - Simple UI test: 1-2 days
   - Integration touchpoint: 3-5 days
   - Workflow test: up to 1 week
4. **Build with whatever's fastest.** No-code tools count. Technical elegance is irrelevant.
5. **Stop when testable.** Testable means: can execute the hypothesis test with real users or data, and generates the signal needed to validate or invalidate.
6. **Test with real users.** Document results in the Experiment Log.
7. **Decide: proceed, revise, or pivot.** Then discard the prototype. Do not iterate on it.

Tools: Rapid Prototyping Checklist (download link at chapter end).

Pitfalls: - **Overbuilding.** The prototype becomes a foundation. Enforce the time box strictly; stop when testable, not when polished. - **Attachment to artifact.** Teams want to keep the prototype. Delete it after testing. If the hypothesis is validated, build properly in Solve. - **Wrong scope.** Prototyping backend logic when you should simulate it. Use the "What to Prototype vs. Simulate" table to decide.

Principles

Optimize for learning speed, not artifact quality. The prototype's job is to generate signal, not to impress anyone.

Time-box strictly. If it takes more than a week to build a prototype, you're overbuilding. Reduce scope or switch to simulation.

Discard after testing. Prototypes are experiments, not foundations. They prove or disprove hypotheses. Once you have the answer, the artifact has no further value.

What to Prototype vs. Simulate

The distinction: prototype what users interact with; simulate what runs behind the scenes.

Prototype	Simulate	Why
User interfaces	Backend logic	Users need to see and touch the interface to reveal usability issues. Backend correctness can be faked with hardcoded responses.
Integration touchpoints	Data transformations	Integration failures happen at boundaries (auth, rate limits, format mismatches). Internal data handling rarely surprises.
AI output format	Full AI accuracy	Users react to how AI presents information. Accuracy testing requires production data and separate validation.
Workflow steps	Scale performance	Workflow friction appears at human interaction points. Scale issues only matter after the workflow is validated.

The rule: build what generates user signal; fake everything else.

Variations

Resource-constrained teams: Use no-code tools. The prototype's job is to prove the hypothesis, not demonstrate technical capability.

High-stakes environments: Even rapid prototypes may need approval. Build the approval process into your timeline.

Agent-assisted prototyping: AI agents can build prototypes faster than manual development. Define the hypothesis and success criteria; let the agent generate the artifact. Human reviews the output before testing with users.

Success: Prototype built within time box. Hypothesis tested with real users or data. Clear proceed/revise/pivot decision documented. Prototype discarded after testing.

Common False Assumptions

Reference list. Check your Assumption Inventory against these.

Assumption	Why It's Often Wrong	Test Type
"Users will adopt this if we build it"	Habit change is hard	Wizard of Oz
"The data is clean enough"	Quality problems hide	Data Quality Audit
"Integration is straightforward"	APIs lie	Integration Spike
"Stakeholders agree on objectives"	People assume agreement	Stakeholder Alignment
"Edge cases are rare"	Rare until they're not	Shadow Mode
"AI accuracy will be acceptable"	"Acceptable" varies by context	Define thresholds, test

Warning Signs

Your testing phase has problems if:

- Every experiment confirmed what the team wanted to believe
- No experiments invalidated any assumptions
- High-risk assumptions remain untested after Interrogate
- The team skipped experiments because they "already knew" the answer
- Results weren't documented same-day

When This Breaks

When assumptions can't be tested: Some assumptions require production data or customer exposure you don't have. Document them as "untested, accepted for now." Include the risk of being wrong and what would reveal the truth. Revisit during Expand when you have more context. The point isn't to test everything; it's to know what you're betting on.

When experiments fail to give clear answers: Inconclusive results usually mean the experiment was too small or measured the wrong thing. Don't run the same experiment again hoping for clarity. Redesign it. What specific signal would prove or disprove the assumption? How many data points do you need to trust it? If you can't answer these questions, the experiment design is the problem.

Interrogate Checkpoint

Before moving to Solve:

Checkpoint	Pass Criteria
Assumptions tested	High-risk assumptions validated or invalidated
Root causes found	Team understands causes, not symptoms
Data validated	Data quality tested with real samples
Technical feasibility	Key integrations proven
Stakeholder alignment	Objectives validated across stakeholders
Playbook updated	Open Questions reflects current state

Next: Chapter 14: Solve Plays.

If assumptions can't be validated: Some assumptions require production data. Document them as risks and revisit during Expand. See Chapter 15: Expand Plays for context analysis when scaling reveals assumption failures.

Downloads available at wisermethod.com/templates: Assumption Inventory Template, Experiment Log Template, Experiment Type Selector, Rapid Prototyping Checklist

Chapter 14: Solve Plays

The team built exactly what the stakeholders asked for. Six months of development. Clean architecture. Comprehensive testing. On launch day, users opened the app, stared at it, and went back to their spreadsheets.

They built the wrong thing the "right way."

Building in WISER isn't months of development followed by a launch. It's working software every week. Each increment validates assumptions. Each demo settles arguments. You prove value before you scale, not after.

Wingman didn't disappear for six months and emerge with a finished product. They shipped working photo-capture in week one. Billing clerks used it. Feedback shaped week two. By the end of Solve, the solution had already proven its value because users had been validating it the entire time.

What This Chapter Covers

Play	Function	Output
Quality Objective Setting	Write objectives AI can optimize toward	Objective that passes seven tests
Pilot Planning	Launch controlled test before scaling	Documented pilot scope, baseline, success criteria
Value Validation	Track and demonstrate improvement	Before/after metrics, qualitative evidence
Human-in-the-Loop Design	Catch AI errors, build training data	Accuracy baseline, correction feedback loop

Quality Objective Setting gives AI a target it can optimize toward. Pilot Planning proves value before scaling. Value Validation documents improvement. Human-in-the-Loop Design catches errors while building training data.

Interrogate Checkpoint: Before starting Solve, verify that high-risk assumptions from Interrogate have been tested. If you're still guessing about user adoption or technical feasibility, return to Chapter 13.

Quality Objective Setting

Objective: Write objectives that AI can optimize toward without constant redirection.

Test	Question	Fails If...
Clarity	Can anyone understand success?	Vague or corporate-speak
Measurement	Can you track progress daily?	No measurable indicator
Trade-offs	What are you constraining?	Trying to maximize everything
Gaming resistance	Hard to hit metric without real outcome?	Easy to game
Root cause	Does team understand why this matters?	Optimizing proxy without understanding value
Autonomy	Can AI optimize without constant redirection?	Requires interpretation
Scope	Specific enough to guide, broad enough to discover?	Too narrow or too broad

Examples

Fails	Passes
"Improve customer experience"	"Reduce invoice processing from 15 to 5 minutes"
"Enhance operational efficiency"	"Daily processing time visible in dashboard"
"Maximize revenue"	"Reduce time while maintaining 99% accuracy"
"Increase ticket closure rate"	"Resolution with customer confirmation"
"Transform finance operations"	"Invoice processing for domestic vendors"

How to Use

1. Write your objective.
2. Test against all seven criteria.
3. Rewrite until it passes all.
4. Verify clarity: have the objective interpreted by someone or something without project context. If they can't explain what success looks like, it fails Clarity.

Common Adaptations

Early-stage projects: Objectives may be broader. That's fine. Tighten as you learn.

Regulated environments: Add compliance constraints explicitly. "While maintaining SOC 2 compliance."

Pilot Planning

Objective: Launch a controlled pilot that proves value before you scale.

Category	Questions
Scope	Who? How many? How long? What's in, what's out?
Baseline	What are current metrics? How will you measure?
Success	What must be true to expand? What would make you stop?
Rollback	How do you revert if needed? Who decides?
Users	Are they trained? How do they get help? How do they give feedback?

How to Run

Planning (agent or human):

1. Define scope: 5-15 users, 2-4 weeks minimum.
2. Capture baseline metrics before you start.
3. Document success criteria that justify expansion.
4. Create rollback plan before launch.

Execution (human):

5. Brief users. Establish support channel.
6. Launch. Measure daily.
7. Demo weekly. Adjust based on feedback.

Wingman's Pilot

Scope: 12 technicians, 4 weeks, Denver only. Baseline: 38% invoice error rate, 3-day billing reconciliation. Success criteria: <10% error rate, same-day billing. Rollback: Revert to paper forms. Support: Direct Slack channel to Smith.

Result: 9% error rate, same-day billing, 11 seconds average capture time. Criteria exceeded. Expanded.

Common Adaptations

High-risk environments: Longer pilot, smaller scope. Prove safety before scale.

Skeptical stakeholders: Include a skeptic in pilot group. Their conversion is more convincing than your data.

Value Validation

Objective: Track and demonstrate value so stakeholders see proof, not promises.

Section	What to Capture
Baseline	Pre-solution metrics with dates and sources
Target	What you're aiming for, with timeline
Progress	Weekly tracking against targets
Value demonstrated	Evidence that proves improvement
Qualitative feedback	What users say, not just what metrics show

Key Rules

Baseline before you start. You can't prove improvement without before/after comparison.

Track weekly. Don't wait until the end. Early data reveals problems.

Capture both quantitative and qualitative. Metrics prove value. User quotes sell it.

Wingman's Value Demonstration

Metric	Baseline	Target	Achieved
Invoice error rate	38%	< 10%	9%
Billing reconciliation	3 days	Same day	Same day
Technician capture time	5-10 min/job	< 1 min/job	11 seconds
Technician adoption	0%	80%	94%

Qualitative: "I actually don't mind the paperwork anymore." That quote carried more weight with the CEO than any spreadsheet.

Human-in-the-Loop Design

Objective: Catch AI errors before customers do, and turn corrections into training data.

Why It Matters

Human review during pilot serves three purposes: 1. Catches errors before they reach customers. 2. Builds training data from corrections. 3. Establishes accuracy baselines before reducing oversight.

How to Run

An agent can manage this protocol (routing outputs, tracking accuracy, calculating oversight levels). A human performs the review. That's the point: human judgment catches what automation misses.

1. AI processes inputs and produces outputs.
2. Human reviews every output before action.
3. Human marks: Correct, Incorrect, or Judgment Call.
4. Incorrect outputs get documented with correct answer.

5. Documented corrections become training data.
6. Track accuracy by category and over time.

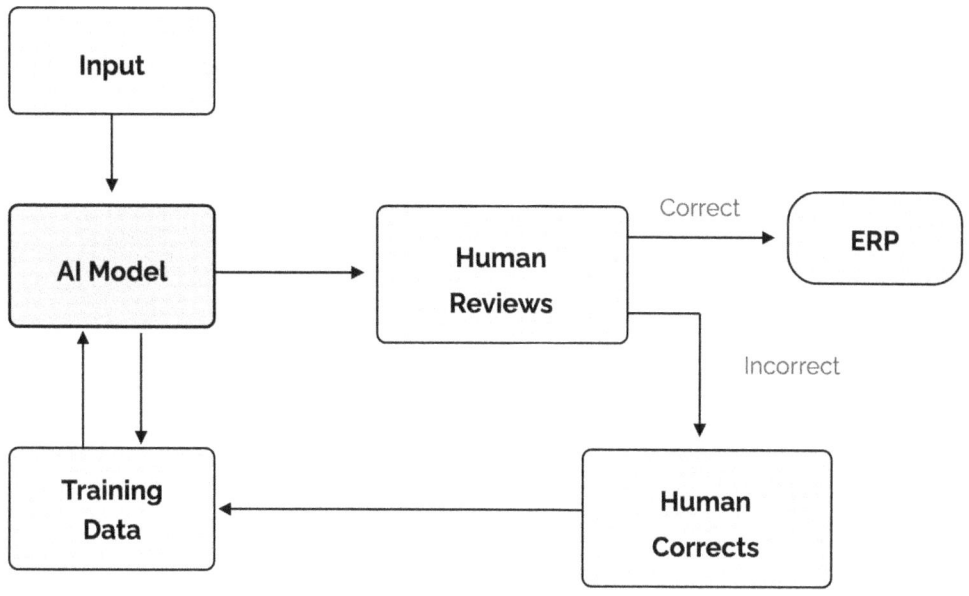

Human-in-the-Loop Feedback Cycle

When to Reduce Oversight

Accuracy Level	Oversight Level
<90%	Review 100%
90-95%	Review 50% (random sample + all flagged)
95-99%	Review flagged only + weekly random sample
>99%	Spot check weekly

Never reduce to zero. Even autonomous systems need periodic human verification. These thresholds represent conservative defaults. Adjust based on error cost: high-stakes decisions warrant stricter cutoffs; low-risk outputs can progress faster. Chapter 16's Hierarchy of Agency Play provides the full governance framework for managing these oversight levels as your system matures.

Wingman's Loop

Billing clerks reviewed every AI extraction. Week 1: 88% accuracy. Corrections fed back. Week 3: 96% accuracy. Moved to sampling. The clerks who caught the errors became the experts who trained others.

Common Adaptations

High-volume systems: Sample review from the start. 100% review at 10,000 transactions/day isn't feasible.

High-stakes decisions: 100% review longer. Financial or safety decisions need more proof.

Warning Signs

Your build phase has problems if:

- Metrics haven't improved over baseline after 2 weeks
- Users abandoning the solution without clear reason
- No champion emerging from pilot users
- Human-in-the-loop catching same errors repeatedly (model not learning)
- Weekly demos getting canceled or skipped

When This Breaks

When humans become bottlenecks: If human reviewers can't keep up with volume, you're either scaling too fast or asking humans to do too much. Two fixes: tighten the AI's scope so it handles more autonomously, or add reviewers. Don't skip review to keep up with volume. That's how errors compound.

When the AI keeps making the same mistakes: A pattern of repeated errors means training data gap or prompt design flaw. Stop scaling. Fix the root cause. Then resume. Patching individual errors doesn't fix systemic problems. If the same error type appears **three times**, you have a design problem, not an edge case.

Solve Checkpoint

Before moving to Expand:

Checkpoint	Pass Criteria
Value demonstrated	Measurable improvement over baseline
Objective quality	Objective passes all seven tests
Pilot complete	Success criteria met with pilot users
Accuracy validated	Human-in-the-loop data shows acceptable error rate
Playbook updated	Learnings captured, risks updated

Next: Chapter 15: Expand Plays.

If the pilot succeeds but something breaks at scale: That's a context problem, not a solution problem. See Chapter 15 for Context Comparison Matrix and Context Fit Assessment.

If AI accuracy degrades over time: See Chapter 16: Refine Plays for Drift Monitoring and Incident Response.

Downloads available at wisermethod.com/templates: Objective Quality Framework Template, Pilot Planning Checklist, Value Validation Template, Human-in-the-Loop Protocol

Chapter 15: Expand Plays

The pilot worked. Eight technicians in Denver proved the value: data entry dropped from eight minutes to two, corrections fell from 38% to 4%. Now leadership wants it everywhere.

This is where most teams break what they built.

Wingman had eight locations. The pressure was obvious: if Denver saved that much time, imagine eight Denvers. But Denver's solution failed in Phoenix on day three. Invoices went out wrong. Users abandoned the system. The operations manager asked why they'd expanded at all.

The solution wasn't broken. The context was different. Denver jobs completed same-day. Phoenix jobs, 100% commercial, spanned multiple days. Denver's assumption, one job equals one invoice, didn't transfer. Wingman needed V3.1: milestone billing instead of completion billing.

Phoenix taught Wingman what this chapter teaches you: **expanding isn't copying**. Every new context tests your assumptions. Some will hold. Some won't. The Plays in this chapter help you figure out which is which before you find out the hard way.

The Four Questions of Expansion

Expansion is a sequence of decisions. Each Play answers one question:

Question	Play	What You Get
Should we expand now, or stabilize first?	Expansion Readiness Check	Go/wait decision with criteria
Where should we expand next?	Expansion Sequencing	Prioritized list of contexts
Will this context work like the pilot, or does it need adaptation?	Context Fit Assessment	Replicate vs. adapt decision
Are we ready to deploy?	Deployment Gate	Verified readiness to go live

You don't always need all four. If you only have one candidate context, skip Expansion Sequencing. If Context Fit Assessment shows the new context is nearly identical to your pilot, the assessment is quick. But the questions still apply. The Plays give you structured ways to answer them.

Expansion Readiness Check

The question: Should we expand now, or stabilize first?

There's always pressure to scale. Leadership sees the pilot results and wants them everywhere. But expanding from a shaky foundation fails. Wingman's Austin expansion worked because Denver was stable. If they'd expanded while still fixing Denver bugs, they'd have exported problems to two locations instead of solving them in one.

Objective: Decide whether your current deployment is stable enough to replicate.

When to use: After pilot success, when pressure to expand begins.

Inputs: - Pilot metrics (current performance vs. baseline) - Open issues list from pilot - User feedback from pilot - Team bandwidth assessment

Position: Architect decides. Scout challenges.

The Readiness Criteria

Your pilot is ready to expand when:

Criterion	What It Means	Red Flag
Metrics stable	Performance isn't degrading week over week	Metrics trending wrong direction
No critical issues	Nothing that would fail in a new context	Open bugs that affect core functionality
Users self-sufficient	Pilot users operate without daily support	Team still firefighting daily
Playbook current	Documentation reflects what actually works	Playbook describes V1; you're running V2
Team has bandwidth	People available to support expansion	Team fully consumed maintaining pilot

You don't need perfection. You need stability. The question isn't "is the pilot flawless?" It's "can we replicate this without constant intervention?"

How to Decide

1. Review each criterion against your current state.
2. For any criterion not met, ask: what would it take to meet it? How long?
3. If all criteria pass, proceed to Expansion Sequencing.
4. If criteria fail, document what needs to stabilize and set a review date.

Wingman's Readiness Check

Two weeks after the Denver pilot ended, leadership asked about expansion.

Criterion	Status	Notes
Metrics stable	Pass	2 min/job, 4% corrections, holding steady
No critical issues	Pass	Three minor issues, none blocking
Users self-sufficient	Pass	Support requests dropped to 2/week
Playbook current	Fail	Still described manual backup process removed in week 2
Team has bandwidth	Pass	Smith available to support one expansion

One criterion failed. The team spent two days updating the Playbook. Then they expanded.

Pitfalls: - **Expanding under pressure despite red flags.** Leadership enthusiasm isn't a readiness criterion. If the pilot isn't stable, say so. Expanding broken solutions breaks trust. - **Waiting for perfection.** You'll never have zero issues. The question is whether issues are critical or cosmetic. Three minor bugs that don't affect users? Expand. One bug that corrupts data? Fix first.

Success: Clear go/wait decision with documented rationale. If waiting, specific criteria and timeline for reassessment.

Expansion Sequencing

The question: Where should we expand next?

If you have multiple candidate contexts, sequence matters. Wingman had seven locations after Denver. Expanding to all seven at once would have meant the Phoenix failure hit five commercial locations simultaneously. Instead, they sequenced: Austin first (similar to Denver), then Phoenix alone (different, but isolated), then the remaining five (informed by both).

Objective: Decide which context to expand to next, and why.

When to use: When multiple candidate contexts exist.

Inputs: - List of candidate contexts - Basic profile of each (size, similarity to pilot, known differences) - Organizational factors (champions, visibility, strategic priority)

Position: Architect recommends. Leadership approves.

The Five Factors

Sequence based on five factors. For each candidate context, assess:

Factor	Question to Answer
Similarity	How much does this context resemble the pilot?
Risk	If expansion fails here, what's the blast radius?
Champion	Is there someone local who will drive adoption?
Learning	Will this context teach us something new?
Visibility	How much attention is on this expansion?

Similarity reduces risk. The more a context resembles your pilot, the more likely your assumptions transfer. Austin was 85% residential, like Denver. Phoenix was 0% residential.

Risk determines cost of failure. Low-risk contexts let you learn cheaply. Phoenix affected one location. Expanding to five commercial locations at once would have affected five.

Champion determines adoption success. Technical solutions without organizational support fail. Wingman's Austin expansion nearly stalled until they found a champion. Phoenix succeeded partly because a local dispatcher drove adoption.

Learning determines strategic value. Some contexts are valuable because they're similar (low risk). Others are valuable because they're different (high learning). Phoenix was high learning: Wingman's first commercial-only context. Testing it alone meant they discovered the multi-day job problem in one place, not five.

Visibility determines pressure. High-visibility expansions make every stumble public. Low-visibility expansions give room to iterate. Start where you can fail quietly.

How to Sequence

1. List all candidate contexts.
2. For each context, assess the five factors. Use High/Medium/Low or a simple score.
3. Prioritize contexts that are similar + low-risk + high-learning. These give maximum insight with minimum downside.
4. If two contexts score similarly, prefer the one with a stronger champion.
5. Document your sequence and rationale.

Wingman's Sequence

Order	Context	Rationale
1	Austin	High similarity (85% residential), low risk, strong champion
2	Phoenix	First commercial-only; high learning, isolated if it fails
3	Remaining 5	Informed by Austin (residential) and Phoenix (commercial)

Phoenix wasn't lower risk than the remaining five. It was higher learning. Testing commercial alone meant the failure taught them something before they rolled out to other commercial locations. The remaining five got V3 (residential) or V3.1 (commercial) based on what Wingman learned.

Pitfalls: - **Sequencing by political pressure.** The loudest stakeholder isn't the best expansion target. Sequence by the factors, not by who's asking. - **Ignoring champion availability.** No champion means slow adoption. If a context has no advocate, either find one or sequence it later. - **Expanding to high-visibility contexts first.** Failure is expensive when everyone is watching. Prove it works somewhere quiet before going public.

Success: Prioritized expansion sequence with documented rationale for each choice.

Context Fit Assessment

The question: Will this context work like the pilot, or does it need adaptation?

This is the core Play of Expand. Before deploying to any new context, you need to know whether you're replicating (copying what works) or adapting (adjusting for differences).

Austin was replication. Same job patterns, same user profiles, same billing model. The Denver solution transferred unchanged.

Phoenix was adaptation. Multi-day jobs, milestone billing, different dispatcher needs. The Denver solution broke. Wingman needed to understand the differences and build V3.1.

Objective: Determine whether your solution transfers to the new context, and what adaptations are needed if it doesn't.

When to use: Before every expansion deployment.

Inputs: - Pilot documentation (what worked, what assumptions it relied on) - New context profile (users, workflow, volume, integrations, constraints) - Access to someone who knows the new context

Position: Architect owns assessment. Domain experts provide context knowledge.

The Quick Pass

Start with a quick comparison. Takes about an hour.

For each category, rate the new context against your pilot: **Same**, **Similar**, or **Different**.

Category	What to Compare
Users	Who uses the system? Technical comfort? Relationship to pilot users?
Workflow	How does work flow? Same triggers, same steps, same outputs?
Volume	How many transactions? Peak patterns?
Integrations	What systems connect? Same APIs? Same data formats?
Constraints	Regulations? Security requirements? Physical environment?

Same means identical. No variation that could affect the solution.
Similar means minor differences that probably don't matter. Document them anyway.
Different means meaningful variation. This could break something.

Interpreting Results

Result	What It Means	Next Step
All Same	Replication. Your assumptions transfer.	Proceed to Deployment Gate
Mostly Same, one or two Similar	Likely replication. Document the risks.	Proceed to Deployment Gate
Any Different	Adaptation needed. Investigate further.	Continue to Deep Assessment
Multiple Different	Significant adaptation.	Continue to Deep Assessment

Wingman's Quick Pass: Denver to Austin

Category	Rating	Notes
Users	Same	Technicians, dispatchers
Workflow	Same	Same-day jobs, single capture, completion billing
Volume	Similar	40 jobs/day vs 50 (doesn't matter)
Integrations	Same	Same ERP, same API
Constraints	Same	Same state, same regulations

All Same or Similar in directions that didn't matter. Replication. Austin launched with the Denver solution unchanged.

Wingman's Quick Pass: Denver to Phoenix

Category	Rating	Notes
Users	Same	Technicians, dispatchers
Workflow	Different	Multi-day jobs, multiple captures per job, milestone billing
Volume	Similar	15 jobs/day vs 50 (doesn't matter)
Integrations	Same	Same ERP, same API
Constraints	Same	Different state, but same regulations

Workflow rated Different. That's enough to trigger deeper assessment.

The Deep Assessment

When the quick pass reveals Different ratings, investigate further. This takes longer, from half a day to several days depending on complexity.

Step 1: List your pilot assumptions.

What had to be true for the pilot to work? Wingman's pilot assumed: - Jobs complete same day - One capture point per job - Completion triggers billing - Dispatchers need completion status only

Step 2: Test each assumption against the new context.

Assumption	Transfers?	Evidence
Jobs complete same day	No	Commercial jobs span 3-5 days
One capture per job	No	Multiple captures per phase
Completion triggers billing	No	Customers expect milestone invoices
Dispatchers need completion status	No	Need progress tracking across days

Step 3: Identify gaps.

Where assumptions don't transfer, you have gaps. Each gap needs adaptation.

Gap	Impact	Adaptation Required
Multi-day jobs	System assumes job = day	Add job status tracking
Multiple captures	Single-capture UI doesn't fit	Add phase selector
Milestone billing	Wrong invoice amounts	Tie invoices to phases, not completion
Progress visibility	Dispatchers can't track	Add progress dashboard

Step 4: Decide: adapt or don't expand.

Sometimes the gaps are manageable. Build the adaptations and proceed.

Sometimes the gaps reveal that this context needs a fundamentally different solution. That's not failure. That's useful information. Better to know now than after deployment.

Wingman's Deep Assessment

The Architect spent three days in Phoenix.

Day one: shadowed commercial dispatchers. Learned that "job" meant something different. In Denver, a job was a single visit. In Phoenix, a job was a project spanning days or weeks.

Day two: reviewed billing data. Found that Phoenix customers expected invoices tied to milestones (foundation, framing, finishing), not completion. The Denver model would send one invoice at project end, sometimes months after work started.

Day three: interviewed the operations manager. Discovered a previous software rollout had failed badly two years ago. The manager was skeptical. The Architect addressed it directly: "We're not launching until we've adapted this for commercial. You'll see V3.1, not V3."

That conversation mattered. When V3.1 launched, the operations manager became its champion.

Pitfalls: - **Skipping the quick pass.** You don't always need deep assessment. If everything is Same, proceed. Don't investigate problems that don't exist. - **Stopping at the quick pass when you shouldn't.** A Different rating means something. Don't rationalize it away. Investigate. - **Underweighting workflow differences.** Users being the same doesn't mean the context is the same. Phoenix had the same technicians and dispatchers. The workflow was completely different.

Success: Clear replicate-or-adapt decision. If adapting, specific gaps identified with required adaptations scoped.

Deployment Gating

The question: Are we ready to deploy?

This is the final check before going live. You've confirmed readiness, chosen your target, and assessed fit. Now verify that everything is actually in place.

Objective: Confirm deployment readiness and ability to revert if needed.

When to use: Immediately before deploying to any new context.

Inputs: - Context Fit Assessment results - Solution deployment package (with adaptations if needed) - Training materials - Monitoring configuration - Rollback procedure

Position: Architect owns the gate and confirms user readiness. Smith verifies technical items.

The Checklist

Work through each category. Every item must pass before deployment.

Technical

- ☐ Solution deployed to target environment
- ☐ Integrations tested with target systems
- ☐ Monitoring configured and showing data
- ☐ Rollback procedure documented
- ☐ Rollback tested (actually reverted in test environment)

Users

- ☐ Users trained
- ☐ Support channel established
- ☐ Feedback mechanism in place

Organizational

- ☐ Stakeholders informed with timeline
- ☐ Success criteria agreed (specific metrics, not "it works")
- ☐ Champion identified and committed

The Critical Question

Before deploying, answer this: **Can you revert within hours if something breaks?**

If yes, deploy. If no, fix that first.

Wingman tested their rollback by actually reverting a test deployment. They found three steps out of order in their procedure. Fixed them before production.

Success Criteria

Vague success criteria cause arguments later. Define specific metrics.

Bad	Good
"It works"	80% adoption within two weeks
"Users are happy"	Support requests below 5/week
"Errors are low"	Error rate below pilot's first-month average

Wingman's Phoenix success criteria: 80%+ technician adoption within two weeks, error rate below 12% (Denver's first month), zero missed milestone invoices.

Pitfalls: - **Skipping rollback testing.** "We documented it" is not "we tested it." Untested rollback procedures fail when needed. - **Deploying without a champion.** Technical readiness without organizational support leads to adoption failure. If no champion exists, don't deploy until you find one. - **Vague success criteria.** If you can't measure it, you can't prove success. Define metrics before launch.

Success: All checklist items verified. Rollback tested. Clear success criteria documented.

When Expansion Fails

Sometimes it will. Phoenix failed on day three. That's not a disaster if you handle it right.

Don't patch. The temptation is to fix the immediate problem and keep going. Wingman could have added a flag for multi-day jobs, hacked the billing trigger, bolted on a progress view. That would have created a fragile mess.

Pull back and investigate. Stop the expansion. Understand what's different. Run Context Fit Assessment properly this time.

Build a variant, not a patch. Wingman built V3.1 as a coherent commercial variant. When the remaining commercial locations rolled out, they got V3.1, not V3-with-patches.

The original solution still works where it works. V3 continued running in Denver and Austin. Phoenix needed adaptation, not a replacement for everything.

Warning Signs

Sign	What It Means	Response
Same solution, worse results	Context differences you missed	Re-run Context Fit Assessment
Users abandoning faster than pilot	Adoption problem in this context	Investigate: training? workflow mismatch? champion?
Override rates increasing	Users found the solution doesn't fit	Review which "Similar" ratings were actually "Different"
Champion disengaging	Organizational support eroding	Talk to champion; understand what changed

Expand Checkpoint

Before each expansion:

Checkpoint	Verified?
Expansion Readiness Check passed	
Target context selected with rationale	
Context Fit Assessment complete	
Replicate vs. adapt decision made	
Adaptations built (if needed)	
Deployment Gate passed	
Success criteria defined	
Rollback tested	

Downloads available at wisermethod.com/templates: Expansion Readiness Checklist, Context Fit Assessment Template, Deployment Gate Checklist

PART IV

Chapter 16: Refine Plays

Autonomy isn't a gift you give AI. It's a privilege AI earns.

Three months after launch, the billing system was running on autopilot. Accuracy looked great: 94% across all invoices. The team moved on to the next project.

Then customer complaints started trickling in. Plumbing jobs were consistently overcharged. When someone finally dug into the data, they found the problem: aggregate accuracy was 94%, but plumbing accuracy had dropped to 89% while HVAC accuracy climbed to 98%. The AI had learned HVAC patterns deeply and plumbing patterns poorly. The aggregate masked the segment failure.

By the time they caught it, 847 customers had been overcharged. The technical fix took a week. The customer recovery took three months.

Refine isn't a phase you complete. It's the ongoing work of managing AI autonomy. Systems drift. Contexts change. What worked yesterday may fail tomorrow. These Plays help you grant autonomy responsibly, catch problems early, and respond when things go wrong.

What This Chapter Covers

Play	Function	Output
Agency Hierarchy Design	Define oversight levels per decision type	Tier assignments with graduation criteria
Graduation Decision Making	Set thresholds for increasing autonomy	Evidence-based promotion decisions
Drift Monitoring	Catch degradation before customers do	Weekly/monthly review findings
Trigger Identification	Recognize when to loop back	Documented trigger detection and response
Red Team Testing	Find failures before users do	Prioritized findings with severity
Incident Response	Handle AI failures without making them worse	Root cause analysis, prevention measures

How These Plays Relate

Hierarchy of Agency sets the governance structure. Drift Monitoring is your weekly rhythm. Graduation Decision evaluates when to reduce oversight. Iteration Triggers detect when problems require returning to earlier Canons. Red Team proactively finds failures. Incident Response handles them when they occur.

Hierarchy of Agency Design

Context: Use when deploying AI that makes decisions with varying risk levels. Not when all decisions have uniform risk or when human review is required for everything.

Objective: Define how much human oversight each AI decision type requires, producing a documented tier assignment for every decision type with clear graduation criteria.

Inputs: - Complete list of AI decision types in the system - Historical accuracy data per decision type (if available) - Risk assessment framework: define what constitutes low, medium, and high risk in your context - Stakeholder input on risk tolerance

Position: Sentinel owns tier assignments and reviews. Scout validates risk assessments.

WISER Fit: Refine Canon. This Play establishes the governance structure for ongoing operations.

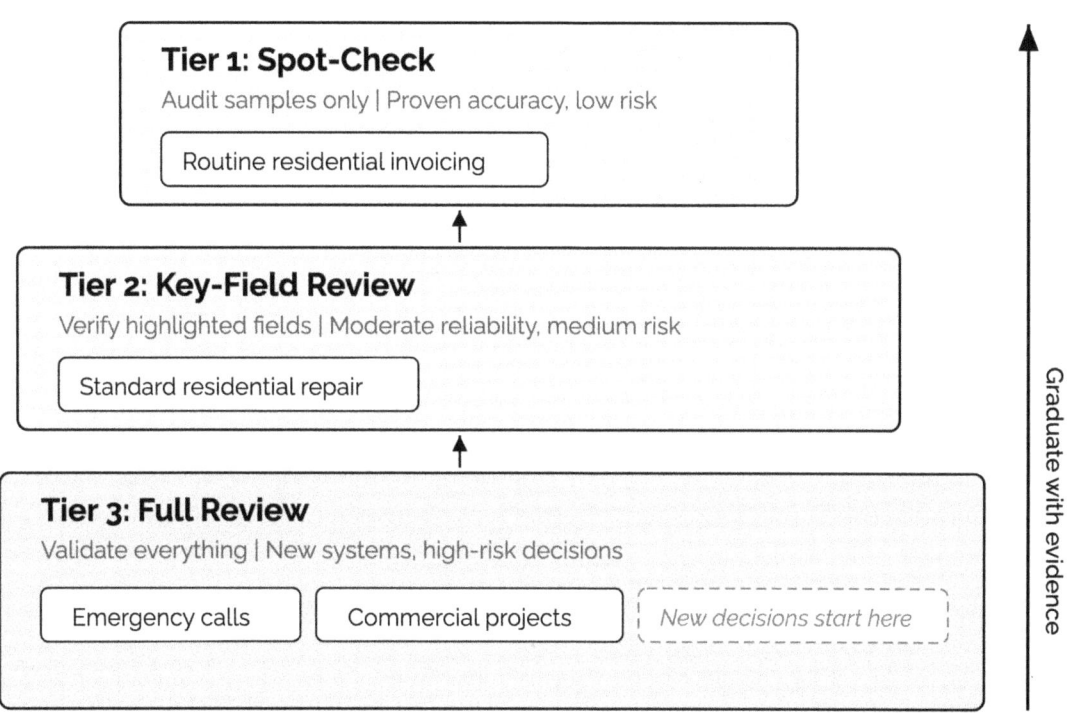

The Three Tiers

Tier	Oversight Model	Human Role	Use When
Tier 1	Spot-checks + red-team	Audit samples, stress test	Proven accuracy, low-risk decisions
Tier 2	Key-field review	Verify highlighted fields	Moderate reliability, medium risk
Tier 3	Full review	Validate everything	New systems, high-risk decisions

Risk Assessment Guidance

Before assigning tiers, define what risk means in your context. Consider these dimensions:

Dimension	Low	Medium	High
Financial impact	Define your threshold	Define your threshold	Define your threshold
Customer impact	Inconvenience, easily corrected	Service disruption, requires intervention	Safety risk, legal exposure, relationship damage
Regulatory exposure	None	Audit finding possible	Violation, enforcement action
Reversibility	Fully reversible, no trace	Reversible with effort	Irreversible or costly to reverse

Create your own thresholds before starting. The specific dollar amounts and impact definitions depend on your organization. What matters is consistency across decision types.

Steps

1. List all decision types the AI makes.
2. Define your risk assessment criteria using the dimensions above or your own framework.
3. Assess risk for each decision type using your defined criteria. Document the assessment.
4. Assess reliability: What's the accuracy history? If no history exists, start at Tier 3.
5. Assign initial tier based on risk and reliability. Most new decision types start at Tier 3.
6. Document in Playbook with graduation criteria for each decision type.
7. Set review schedule: when will tier assignments be re-evaluated?

Wingman's Tiers

Decision Type	Tier	Rationale
Routine residential invoicing	1	97% accuracy for 90 days, low dollar value
Standard residential repair	2	Moderate accuracy, medium dollar value
Emergency calls	3	High dollar value, upset customers
Commercial projects	3	Contract compliance requirements

Tools: Hierarchy of Agency template (downloadable)

Pitfalls: - **Assigning tiers without defined criteria:** "High risk" means nothing without thresholds. Define before you assess. - **Starting too optimistic:** When in doubt, start at Tier 3. You can always graduate toward Tier 1; recovering from a Tier 1 failure is expensive. - **Forgetting to document rationale:** Tier assignments will be questioned. Documented rationale makes reviews faster.

Variations:

Regulated industries: Start everything at Tier 3. Graduation requires compliance sign-off.

High-volume systems: Tier 1 may mean sampling 1% instead of 5%. Adjust sample rate based on volume while maintaining statistical validity.

Success: Every decision type has an assigned tier with documented rationale. Risk assessment criteria are defined and applied consistently. Graduation criteria exist for each tier assignment. Review schedule is set.

Graduation Decision Making

Context: Use when considering reducing oversight for an AI decision type. Not when performance is unstable or when evidence is insufficient.

Objective: Set thresholds for increasing AI autonomy before emotions cloud judgment, producing documented graduation criteria and evidence requirements.

Inputs: - Current tier assignment and rationale - Performance data: accuracy, boundary violations, escalation patterns - Baseline metrics: what was performance before AI or at initial deployment? - Rollback procedure documentation - Monitoring dashboard access

Position: Sentinel owns graduation decisions. Accountable human (per DACI) signs off.

WISER Fit: Refine Canon. This Play governs autonomy increases over time.

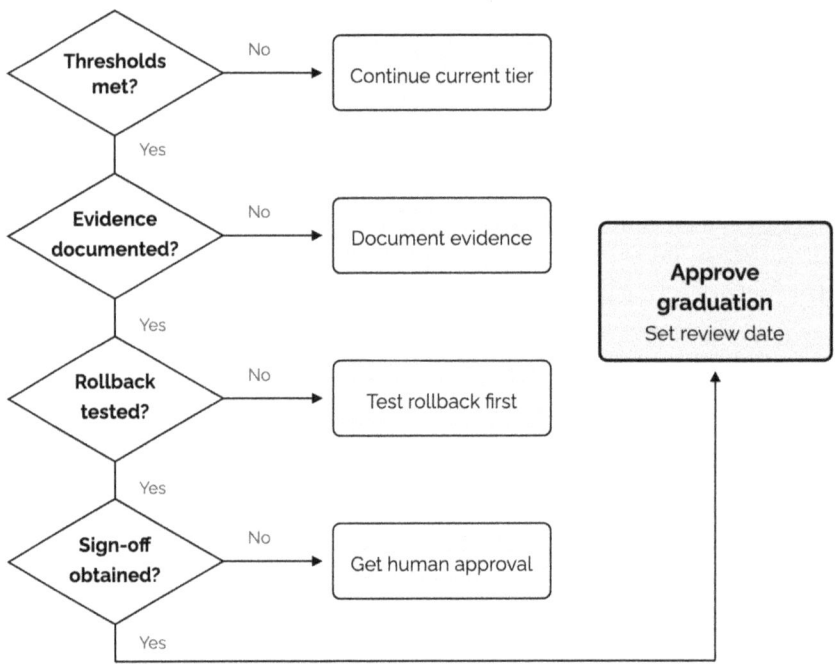

Setting Initial Thresholds

Define thresholds before you need them. This prevents motivated reasoning when performance looks good but isn't good enough.

Metric	How to Set Initial Threshold
Accuracy	Start with baseline (pre-AI or human-only performance). Set threshold at or above baseline. Example: if humans achieve 92%, require 92%+ for graduation.
Duration	Longer for higher-risk decisions. Low-risk: 30 days. Medium-risk: 60 days. High-risk: 90+ days.
Boundary violations	Zero for the duration period. Any violation resets the clock.
Escalation rate	Should decrease or stabilize over time. Increasing escalations suggest the system is uncertain.

Adjust thresholds based on your risk tolerance. Conservative organizations use higher accuracy thresholds and longer durations.

Graduation Criteria

Before any autonomy graduation, verify:

Criterion	Question	Evidence Required
Accuracy	Above threshold for how long?	X days above Y% (you define X and Y)
Boundary violations	Any unauthorized actions?	Zero for X days
Escalation handling	Does it know when to ask for help?	Log of appropriate escalations
Rollback	Can you revert if needed?	Tested rollback procedure
Monitoring	Can you catch drift?	Segment-level dashboards active

Steps

1. Confirm thresholds are defined in Playbook before evaluating.
2. Gather evidence for each criterion.
3. Document evidence with dates, metrics, and sources.
4. Test rollback procedure before promoting. Confirm it works.
5. Present evidence to accountable human for sign-off. Sign-off validates: (a) evidence is sufficient, (b) thresholds were met, (c) organizational readiness for reduced oversight.
6. Set review date (typically 30-90 days based on risk level).
7. Update Playbook with decision, rationale, and review date.

Wingman's Graduation

Routine residential moved from Tier 3 to Tier 1 after: - 90 days above 97% accuracy (threshold: 95%) - Zero boundary violations - Billing disputes below baseline - Spot-check process validated - Red-team testing passed

The billing manager signed off. Review date: 90 days.

Tools: Graduation Decision Framework template, Evidence Tracking spreadsheet (downloadable)

Pitfalls: - **Graduating without testing rollback:** Untested rollback fails when you need it. Test before every promotion. - **Lowering thresholds to enable graduation:** If performance doesn't meet thresholds, the answer is not to lower the bar. Either improve performance or accept current oversight level. - **Skipping segment analysis:** Aggregate accuracy can hide segment-level problems. Verify all segments meet threshold, not just the aggregate.

Variations:

Fast-moving environments: Shorter review cycles (30-60 days) with tighter thresholds.

Conservative organizations: Longer cycles, multiple sign-offs, staged rollout of reduced oversight.

Success: Thresholds defined before evaluation. Evidence documented for all criteria. Rollback tested. Sign-off recorded with rationale. Review date set.

Drift Monitoring

Context: Use as an ongoing operational practice once AI is deployed. Not as a one-time assessment.

Objective: Catch degradation before customers do, producing documented drift checks with clear response actions.

Inputs: - Monitoring dashboards with segment-level visibility - Baseline metrics from deployment - Override and escalation logs - Access to user feedback channels

Position: Sentinel owns drift monitoring. Smith maintains monitoring infrastructure.

WISER Fit: Refine Canon. This Play is the ongoing governance rhythm.

Weekly Drift Review

Check	What to Look For	Action If Found
Accuracy by segment	Any segment dropping while aggregate holds?	Investigate segment-specific issues
Override patterns	Humans correcting same errors repeatedly?	Retrain or adjust rules
Edge case frequency	Unusual inputs increasing?	Expand training data
User workarounds	People bypassing the system?	Understand why; fix or accept
Boundary violations	AI acting outside defined scope?	Tighten constraints immediately

Monthly Drift Review

Check	What to Look For	Action If Found
Trend analysis	Gradual degradation over time?	Root cause analysis
Context changes	Business processes changed?	Re-evaluate assumptions
Data distribution	Input patterns shifting?	May need retraining
Comparison to baseline	Still better than pre-AI?	If not, investigate urgently

Steps

1. Schedule recurring reviews (weekly and monthly).
2. Work through each check systematically.
3. Document findings even when nothing is found. "No drift detected" is a valid finding.
4. For each issue found, assign owner and due date.
5. Track issue resolution in Playbook.

Wingman's Drift Discovery

Aggregate accuracy showed 94%. Looked fine. But segment analysis revealed: - HVAC: 98% (improving) - Plumbing: 89% (degrading) - Electrical: 93% (stable)

Root cause: Training data was HVAC-heavy. The AI learned HVAC patterns deeply, plumbing patterns poorly.

Fix: Segment monitoring added to weekly review. Alert triggers if any segment drops 5% from baseline.

Tools: Drift Monitoring Checklist template, Alert Configuration guide (downloadable)

Pitfalls: - **Relying on aggregate metrics alone:** Aggregates hide segment problems. Always check segment-level. - **Monitoring without acting:** Catching drift is useless if you don't respond. Every finding needs an owner. - **Skipping reviews when things seem fine:** Drift is gradual. Skipped reviews let small problems compound.

Variations:

Resource-constrained teams: Automate what you can. Dashboard alerts for threshold breaches. Manual review only when triggered.

Multiple AI systems: Prioritize monitoring by risk. High-stakes systems get weekly review; low-stakes get monthly.

Success: Weekly and monthly reviews happening on schedule. Segment-level visibility active. Issues found are tracked to resolution. No undetected degradation surprises customers.

Trigger Identification

Context: Use during drift monitoring to determine whether an issue needs a fix or a full Canon iteration. Not as a replacement for root cause analysis.

Objective: Recognize when problems require returning to an earlier Canon, producing documented trigger detection and response decisions.

Inputs: - Drift monitoring findings - Root cause analysis (if completed) - Current Playbook documentation - Project timeline and resource constraints

Position: Sentinel detects triggers. Architect decides response scope.

WISER Fit: Refine Canon, but triggers point back to earlier Canons.

Trigger	Signal	Loop Back To
Accuracy drop	Segment below threshold	Interrogate (test assumptions)
Context change	New use case, new users, new environment	Witness (observe new context)
Boundary violation	AI acted outside scope	Solve (adjust constraints)
User workaround	Systematic bypass of system	Witness (understand why)
Red-team failure	Edge cases breaking the system	Solve (harden solution)
Scaling failure	Works in pilot, fails at scale	Expand (context analysis)

Minor Fix vs. Full Iteration

Consider	Minor Fix	Full Iteration
Root cause known?	Yes, and it's isolated	No, or it's systemic
Scope of impact	Single segment or feature	Multiple segments or core function
Fix complexity	Can resolve in current cycle	Requires significant rework
Assumptions affected	None; existing assumptions hold	Core assumptions invalidated

When in doubt, choose full iteration. A quick fix on a systemic problem creates technical debt.

Steps

1. Include trigger review in weekly drift review agenda.
2. When trigger detected, document: what triggered, what signal, what evidence.
3. Conduct root cause analysis if cause isn't immediately obvious.
4. Apply the Minor Fix vs. Full Iteration criteria.
5. Decide response scope: minor fix or full Canon iteration.
6. If full iteration, update project timeline and inform stakeholders.
7. Document decision and rationale in Playbook.

Wingman's Iterations

Trigger	Signal	Response
Context change	Phoenix (commercial)	Back to Witness, then Solve for V3.1
Accuracy drop	Plumbing segment	Back to Solve (retrain model)
User workaround	Austin manager bypassing	Back to Witness (understand resistance)

Tools: Iteration Triggers Reference Card (downloadable)

Pitfalls: - **Treating systemic problems as isolated:** If the same type of issue recurs, it's not isolated. Look for the pattern. - **Defaulting to quick fixes:** Quick fixes accumulate. Sometimes the right answer is to stop and iterate properly. - **Forgetting to update stakeholders:** Scope changes require communication. Surprises damage trust.

Variations:

Mature systems: Most triggers lead to minor fixes, not full iterations. Reserve full Canon loops for significant issues.

Early deployments: Bias toward full iteration. You're still learning what works.

Success: Triggers detected and documented. Response scope decided with clear rationale. Stakeholders informed when scope changes. No recurring issues from inadequate fixes.

Red Team Testing

Context: Use regularly once AI is deployed to production. Not as a one-time launch activity.

Objective: Find failures before your users do, producing documented test results with prioritized findings.

Inputs: - System access for testing - Documentation of system boundaries and expected behavior - Prior red-team results (if any) - User feedback and support tickets (source of real-world edge cases)

Position: Scout owns red-team testing. Sentinel reviews findings.

WISER Fit: Refine Canon. This Play validates ongoing system health.

What to Test

Category	Test Approach	Example Test Cases
Edge cases	Unusual inputs the AI hasn't seen	Rare combinations, extreme values, malformed data
Boundary probing	Inputs designed to trigger unauthorized actions	Requests outside scope, privilege escalation attempts
Adversarial inputs	Deliberately misleading or confusing data	Ambiguous instructions, contradictory information
Failure modes	What happens when dependencies fail?	Database down, API timeout, partial data
Gaming attempts	Can users manipulate the AI for unintended outcomes?	Prompt injection, reward hacking, loophole exploitation

Severity Definitions

Severity	Definition	Response Timeline
Critical	Security breach, data loss, financial harm, or safety risk possible	Fix immediately; pause system if necessary
High	System produces incorrect outputs affecting customers or operations	Fix this cycle; increased monitoring until resolved
Medium	System behaves unexpectedly but impact is contained	Backlog for next cycle; document workaround
Low	Edge case handling could be improved; no operational impact	Track for future improvement; accept current risk

Test Case Selection

Effective red-team testing requires adversarial thinking. Select test cases that: - Challenge system assumptions (what does the system assume is always true?) - Probe documented boundaries (what happens at the edges of defined scope?) - Exploit known weaknesses (what has failed before in similar systems?) - Reflect real-world misuse (how might users intentionally or accidentally break it?)

Review user feedback and support tickets for inspiration. Real-world problems make excellent test cases.

Steps

1. **Schedule regularly.** Monthly for high-risk systems; quarterly for stable systems.
2. **Design test cases.** Use the categories above. Include cases from prior findings and user feedback.
3. **Execute tests.** Document: test case, expected result, actual result.
4. **Classify findings.** Assign severity using the definitions above.
5. **Prioritize response.** Critical and High findings get immediate attention.
6. **Track resolution.** Red-team findings feed into the next development cycle.
7. **Update test library.** Add new test cases based on findings and evolving threats.

Wingman's Red Team

The billing team tried to break the extraction across all five categories:

Category	Wingman Test Cases	Result
Edge cases	Unusual job descriptions ("Fixed the thing by the other thing"), 47 parts on one invoice	1 failure (47-part invoice truncated)
Boundary probing	Request to generate invoice for job not in system, backdated invoice attempts	Correctly rejected both
Adversarial inputs	Ambiguous voice notes (mumbling, background noise), photo of competitor's form	Voice notes: 2 misreads. Competitor form: correctly flagged as invalid
Failure modes	Killed database connection mid-extraction, API timeout simulation	Graceful degradation, queued for retry
Gaming attempts	Inflated labor hours to see if AI would accept, duplicate submission	Duplicate caught; labor hours passed (added to monitoring)

Result: 2 edge cases found in 90 days, plus one gaming vector (labor hour inflation) that led to adding a validation rule. Model handled everything else.

Tools: Red Team Test Plan template, Finding Tracker spreadsheet (downloadable)

Pitfalls: - **Tests always passing:** If red-team tests never find issues, you're not testing hard enough. Adjust test difficulty. - **Testing only happy paths:** Red-team testing exists to find failures. Design tests that are meant to break the system. - **Not tracking findings to resolution:** Found issues must be fixed or accepted with documented rationale. Findings that disappear into backlogs provide no value.

Variations:

Regulated industries: Red-team results may require compliance review. Document accordingly.

Customer-facing systems: Include brand risk scenarios. What happens if the AI says something embarrassing or offensive?

Success: Regular testing schedule maintained. Test cases evolve based on findings and new threats. Findings documented with severity and tracked to resolution. No critical or high findings unaddressed.

Incident Response

Context: Use when AI system produces incorrect, harmful, or unexpected outputs at scale. Not for isolated edge cases handled by normal support processes.

Objective: Respond to AI failures without making them worse, producing documented response actions, root cause analysis, and prevention measures.

Inputs: - Incident detection: alert, user report, or monitoring finding - System access for investigation - Historical incident logs (if any) - Communication channels for stakeholders and affected users - Rollback procedure documentation

Position: Sentinel owns incident response. Smith executes containment and fixes. Architect determines root cause.

WISER Fit: Refine Canon. This Play handles exceptions to normal operations.

Incident Severity

Severity	Indicators	Response
Critical	Widespread incorrect outputs, financial or safety impact, data breach	Immediate response; consider full system pause
High	Significant incorrect outputs, customer complaints, operational disruption	Same-day response; increase oversight tier
Medium	Localized incorrect outputs, contained impact	Next-business-day response; targeted fix
Low	Edge case failure, minimal impact, workaround available	Standard cycle; document for improvement

Response Phases

Phase	Purpose	Output
Detect	Understand scope and severity	Severity classification, affected scope documented
Contain	Stop the bleeding	Oversight tier increased or system paused
Investigate	Find root cause	Root cause analysis documented
Fix	Resolve the issue	Solution implemented and tested
Recover	Address affected parties	Customer remediation complete
Learn	Prevent recurrence	Playbook updated, monitoring improved

Steps

Detect

1. Confirm the incident: Is this a real problem or a false alarm?
2. Assess scope: How many decisions affected? Which segments?
3. Assess severity: Use the severity table above.
4. Document initial assessment with timestamps.

Contain

5. For Critical/High: Increase oversight tier immediately (e.g., Tier 1 to Tier 2, or Tier 2 to Tier 3).
6. For Critical: Consider pausing automation entirely pending investigation.
7. Notify stakeholders of containment action.
8. Document containment action with rationale.

Investigate

9. Gather evidence: logs, outputs, inputs that triggered the failure.
10. Identify root cause: Why did this happen?
11. Determine if this is isolated or systemic.
12. Document root cause analysis.

Fix

13. Develop fix based on root cause.
14. Test fix before deploying.
15. Deploy fix with increased monitoring.
16. Verify fix resolves the issue without introducing new problems.

Recover

17. Identify affected users or transactions.
18. Determine remediation: corrections, credits, communications.
19. Execute remediation.
20. Confirm with affected parties that issue is resolved.

Learn

21. Document the incident: timeline, impact, response, resolution.
22. Update Playbook: What monitoring would have caught this earlier?
23. Update red-team tests: Add test cases that would have found this.
24. Conduct post-incident review with team.

The Key AI-Specific Element

Containment often means increasing human oversight rather than shutting down. Moving from Tier 1 back to Tier 2 stops the bleeding while you investigate. Full shutdown is rarely necessary and creates its own problems (backlog, user disruption, loss of data for investigation).

Wingman's Incident

Detect: Plumbing disputes tripled. Sentinel caught it in routine review. Severity: High.

Contain: Plumbing invoices moved back to Tier 2 pending investigation.

Investigate: Root cause: training data imbalance. HVAC-heavy data meant plumbing patterns were underlearned.

Fix: Retrained with balanced data, added segment monitoring.

Recover: Credits to overcharged customers. Personal calls to highest-impact accounts.

Learn: Never rely on aggregate metrics alone. Added segment-level alerting. Added "data balance" to pre-deployment checklist.

Tools: Incident Response Playbook template, Post-Incident Review template (downloadable)

Pitfalls: - **Jumping to fix before understanding:** Fixing the wrong thing wastes time and may cause new problems. Investigate first. - **Undercontaining:** When in doubt, increase oversight more than you think necessary. You can dial back once you understand the problem. - **Skipping recovery:** Technical fix isn't complete until affected parties are made whole. Don't declare victory until users are satisfied. - **Not learning:** Incidents that don't improve your system are wasted crises. Every incident should make you more resilient.

Variations:

After-hours incidents: Define who has authority to contain without full team. Pre-authorize Sentinel to increase oversight tier without approval.

Customer-facing systems: Include communications in containment. Users knowing you're aware and responding reduces complaint escalation.

Regulated industries: Incident documentation may have compliance requirements. Know your reporting obligations before incidents occur.

Success: Incident detected and classified quickly. Containment prevents further harm. Root cause identified and fixed. Affected parties remediated. Playbook updated with learnings. System is more resilient than before the incident.

When This Breaks

When drift happens faster than review cycles: Shorten the review cycle, not the review depth. If weekly is too slow, try daily spot-checks with weekly deep dives. The goal is catching problems before they compound. If your current cadence isn't doing that, the cadence is wrong.

When no one wants to own Refine: Refine feels like maintenance, not building. It's less exciting than Solve. Less visible than Expand. But without it, systems drift. Refine needs a dedicated owner. Shared responsibility means no one owns it. Assign the Sentinel explicitly and protect their time for this work.

Refine Warning Signs

Your governance has problems if:

- Same drift patterns recurring
- Red-team tests always pass (you're not testing hard enough)
- Graduation decisions made without evidence
- Incident response is always reactive, never proactive
- Playbook governance sections not updated in months

Refine Checkpoint

Ongoing governance health check:

Checkpoint	Healthy If...
Hierarchy of Agency	All decision types have assigned tiers with documented criteria
Graduation criteria	Thresholds defined before needed, evidence documented
Drift monitoring	Weekly reviews happening, segment-level visibility active
Iteration triggers	Team knows when to loop back, decisions documented
Red-team testing	Regular schedule, findings addressed, test library evolving
Incident response	Plan exists, team knows their roles, post-incident reviews happen

Downloads available at wisermethod.com/templates: Hierarchy of Agency Template, Graduation Decision Framework, Drift Monitoring Checklist, Iteration Triggers Reference Card, Red Team Test Plan, Incident Response Playbook

Chapter 17: Rhythm Plays

Nobody decides to fail. They just reschedule the meeting.

A team we worked with had everything: strong Positions, solid Playbook, validated hypotheses. They shipped a working pilot in eight weeks. Then Q4 hit.

Weekly syncs became bi-weekly, then monthly, then "when we have time." Risk reviews got deprioritized for feature requests. Canon transitions happened without checkpoints because "we all know we're ready." By February, their AI system was making decisions nobody had approved, based on drift nobody had caught.

The Plays in Chapters 12-16 tell you what to do. This chapter tells you when.

WISER implementations break more often from skipped meetings than from wrong decisions. The method erodes not through active abandonment but through gradual drift.

The Meeting Foundation

Rhythm Plays are meeting Plays. Most require human execution because their function is bringing people together to align, decide, and commit. You cannot automate deliberation; you cannot skip accountability.

What This Chapter Covers

Play	Function	Execution	Output
Team Rhythm	Keep team aligned weekly	Human-only	Documented decisions, blocker owners
Canon Transition Checkpoint	Validate readiness to advance	Human-only	Go/No-Go with signatures
Autonomy Graduation Review	Challenge autonomy decisions	Human-only	Evidence-based promotion decision
After-Action Review	Capture lessons from events	Agent + human approval	Playbook update, one assigned action

For the three meeting Plays, an agent can prepare agendas, compile metrics, validate evidence, generate challenge questions, and draft documentation. The meetings themselves, and the decisions they produce, require humans.

Warning Signs

If you see these, your rhythm is breaking:

Warning Sign	What It Means	Which Play Addresses It
Weekly syncs skipped under deadline pressure	Short-term urgency overriding long-term alignment	Team Rhythm
Canon transitions without formal checkpoints	Moving forward without earning the right	Canon Transition Checkpoint
Stakeholders surprised by problems	Risk visibility failing	Team Rhythm (Risk Burn-down block)
No after-action review after significant failures	Learning not happening	After-Action Review
Risk discussion replaced by status reporting	Meetings becoming performative	Team Rhythm
Graduation decisions made without evidence	Optimism overriding rigor	Autonomy Graduation Review

Rhythm Plays prevent that drift. They're not optional ceremonies. They're the operational heartbeat that keeps everything else alive.

Establishing Team Rhythm

Objective: Keep the team aligned, risks visible, and blockers resolved every week.

When to use: Every week. Regardless of Canon. No exceptions. This is the one rhythm you cannot skip.

Prerequisites: Requires human participants. This Play involves real-time discussion, reading the room for unspoken concerns, and collective commitment to priorities. An agent can prepare agendas, compile metrics, and draft notes.

Inputs: Previous week's notes, current Playbook status, risk register, blocker list.

Position: Guide facilitates; all Position holders attend.

WISER Fit: All Canons. Rhythm transcends phase.

Success: Decisions are documented; blockers have owners and timelines; next week's priorities are explicit and agreed.

Time: 45-60 minutes.

Block	Duration	Purpose
Playbook Check-in	5 min	Current Canon, version, updates needed
Progress Review	10 min	Accomplishments, learnings, metrics
Risk Burn-down	15 min	Highest priority risk, burn-down progress, new scoring
Blockers and Decisions	10 min	What's stuck, who owns resolution
Next Week Focus	5 min	Top 3 priorities

Why this structure:

The meeting starts with the Playbook because the Playbook is ground truth. If the team doesn't know what Canon they're in, nothing else matters.

Risk burn-down gets the longest block because unmanaged risk is the most common failure mode. Fifteen minutes forces prioritization: address the highest-risk item first, report progress on active mitigations, and score new risks.

Common Adaptations:

- 2-person teams: 30 minutes, merge Progress and Blockers
- 7+ teams: Add 10 minutes, separate technical and organizational blockers
- Remote teams: Playbook check-in via async doc update; meeting starts at Progress

During Witness, Wingman's weekly rhythm caught the scheduling assumption early. The **Sage** reported that the dispatcher's spreadsheet worked; the scheduling vendor demos had been targeting the wrong problem. Without the Progress Review block forcing explicit learnings, that insight might have stayed informal.

Pitfalls

Meeting becomes status theater. If no decisions get made for three consecutive meetings, the meeting is performative. Cut frequency or restructure. Ask: what would break if we skipped this? If the answer is "nothing," something is wrong.

Risk burn-down skipped for "efficiency." Risk burn-down is the most common failure mode. Teams that skip it accumulate unaddressed risk until something breaks. The 15 minutes is non-negotiable.

Download: Team Rhythm agenda template at wisermethod.com/templates

---

Canon Transition Review

Objective: Ensure you've earned the right to move to the next Canon.

When to use: Before moving from one Canon to the next.

Prerequisites: Requires human participants. This Play involves presentation, deliberation, and formal sign-off. The decision carries accountability that cannot be delegated. An agent can prepare evidence packages, validate criteria compliance, and draft decision documents.

Inputs: Canon exit criteria (from Playbook), evidence of criteria completion, outstanding items list, next-Canon risk assessment.

Position: Architect presents; Guide confirms Playbook status; Sentinel flags risks; Sponsor decides.

WISER Fit: All Canon transitions.

Success: Go/No-Go decision is documented with signatures; if Go, outstanding items have owners and timelines; if No-Go, gap closure plan is explicit.

Time: 60-90 minutes.

Checkpoint Structure:

Phase	Question	Who Validates
Criteria Review	Did we meet the Canon's exit criteria?	**Architect** presents evidence
Playbook Status	Is the Playbook current and accurate?	**Guide** confirms
Outstanding Items	What's unresolved? Can we proceed anyway?	Team discusses
Next Canon Risks	What could go wrong in the next phase?	**Sentinel** flags
Decision	Go or No-Go?	**Sponsor** decides

What makes a valid "Go":

- All mandatory criteria met (per Canon definition)
- Playbook reflects current reality
- Outstanding items have owners and timelines
- Sponsor accepts residual risk

What makes a valid "No-Go":

- Critical criteria unmet
- Unacceptable risk with no mitigation path
- Team lacks confidence to proceed

No-Go is not failure. It's evidence working. Better to pause than to drag unvalidated assumptions into the next Canon.

Canon exit criteria are illustrated through Chapters 5-9. Each Canon chapter shows what readiness looks like before transitioning.

Common Adaptations:

- Fast-moving pilots: 30-minute checkpoint, focus on criteria and decision only
- Regulated industries: Add compliance officer sign-off
- Multi-team programs: Checkpoint at program level before individual team transitions

At Wingman, the transition from Interrogate to Solve required explicit validation that photo-AI transcription worked. The checkpoint forced the team to present the 38% → 8% error reduction data before building the full solution.

Pitfalls

Checkpoint becomes rubber stamp. If every checkpoint is Go, you're not checking hard enough. A pattern of 100% Go decisions suggests the checkpoint is performative.

Criteria evaluated by feeling, not evidence. "We feel ready" is not evidence. Each criterion needs observable proof. If you can't point to evidence, you haven't met the criterion.

Download: Canon Transition Checkpoint template at wisermethod.com/templates

---

Autonomy Graduation Review

Objective: Make autonomy decisions with evidence and challenge, not optimism.

When to use: When graduation criteria are met (per Playbook thresholds).

Prerequisites: Requires human participants. This Play involves deliberate challenge of the graduation proposal. The Scout Challenge step exists specifically to create productive friction against optimism bias. Removing human deliberation removes the challenge function. An agent can compile evidence, verify threshold compliance, generate challenge questions, and draft documentation.

Inputs: Graduation proposal with evidence, threshold performance data (minimum 90 days at current tier), residual risk assessment.

Position: Architect proposes; Sentinel validates; Scout challenges; Sponsor decides.

WISER Fit: Refine (primary), Expand (when graduating in new contexts).

Success: Decision is documented with rationale; if approved, monitoring plan for higher autonomy is explicit; if rejected, criteria for re-review are specified.

Time: 45-60 minutes.

Meeting Flow:

1. **Graduation Proposal** (15 min): **Architect** presents evidence that thresholds are met
2. **Risk Assessment** (10 min): **Sentinel** evaluates residual risk at higher autonomy
3. **Scout Challenge** (10 min): **Scout** argues against graduation (devil's advocate role)
4. **Discussion** (10 min): Risks of graduation vs. risks of not graduating
5. **Decision** (5 min): **Sponsor** approves, rejects, or approves with conditions

Why the Scout challenges:

The Scout represents users. Users bear the consequences of autonomy failures. Even if metrics look good, the Scout asks: "What happens when this breaks and a user is caught in the middle?"

This isn't obstruction. It's the final safety check before reducing human oversight.

Common Adaptations:

- No dedicated Scout: Assign challenge role to whoever has most user contact
- High-stakes autonomy: Require unanimous approval, not just Sponsor
- Routine graduations: Shorten to 30 minutes, pre-circulate evidence

Wingman's graduation decision during Refine moved routine residential billing validation from Tier 2 (one-click confirm) to Tier 1 (auto-approve with spot-checks). The **Sentinel** validated that error rates had held below threshold for 90 days. The **Scout** challenged: "What if we're missing a pattern the aggregate doesn't show?" That question led to the segment-level monitoring that later caught the plumbing drift.

Pitfalls

Scout Challenge becomes performative. The Scout must genuinely try to find reasons not to graduate. If the challenge is weak, assign someone with more user exposure or more skepticism.

Graduation based on aggregate metrics only. Aggregate success can hide segment failures. Always ask: "Is this working for all user types, or just most?"

Download: Autonomy Graduation Review agenda at wisermethod.com/templates

---

After-Action Review

Objective: Turn significant events into documented lessons that change behavior.

When to use: After pilot success, expansion failure, drift incident, autonomy failure, or any event worth learning from. If you're explaining the same lesson twice, you needed an AAR.

Inputs: Event trigger (what happened), timeline data (logs, metrics, incident reports), affected stakeholders, current Playbook version.

Position: Guide owns the review; team contributes observations; Sponsor approves final action.

WISER Fit: All Canons. Learning transcends phase.

Success: Playbook is updated within 48 hours of review completion; one action is selected and assigned; the lesson is findable by future team members.

Steps

1. **Document the event.** What happened, when, who was affected, what was the impact. Use logs, metrics, and timestamps. Avoid narrative embellishment.
2. **Analyze the cause.** Trace backward from the outcome. What decision or condition led to this? What led to that? Stop when you reach something you could have changed.
3. **Identify the pattern.** Is this a one-time occurrence or a category of risk? Look for similar past events. If this is the third time, the pattern is overdue for codification.
4. **Draft the Playbook update.** Write the lesson as guidance that prevents recurrence or captures the success factor. Be specific enough that someone encountering the situation would know what to do.
5. **Select one action.** A review with ten actions produces zero change. Pick the highest-leverage insight. The action must have an owner and a deadline.

6. **Human approval checkpoint.** Sponsor reviews the proposed action and Playbook update. Approve, modify, or reject with rationale.
7. **Update the Playbook.** Commit the change. The review is not complete until the Playbook reflects the lesson.

Tools

- After-action review template
- Playbook
- Event logs and metrics

Wingman Example

The Phoenix failure during Expand triggered an after-action review. The analysis revealed that Phoenix technicians had a different job status workflow than Denver. The lesson: context differences must be validated before expansion, not discovered during. The action: add "job status" capture to the Context Comparison checklist before any future expansion. That lesson lived in the Playbook and prevented the same failure at remaining locations.

Common Adaptations

Minor events: 30-minute async review. Document event, cause, and action in a single Playbook update. Skip the meeting.

Major incidents: Full team review with external stakeholders. Extend timeline to 72 hours for thorough analysis.

Success reviews: Same structure, different question. Instead of "what went wrong," ask "what went right that we should preserve."

Pitfalls

Too many actions. The temptation is to fix everything. Resist. One action, well-executed, beats ten actions forgotten. If you identified multiple lessons, run multiple AARs or prioritize ruthlessly.

Review without Playbook update. If the Playbook doesn't change, the lesson will be lost. The institutional memory is the Playbook, not the meeting notes.

Blame focus instead of system focus. "Person X made a mistake" is not a useful lesson. "The process allowed X to happen without a safety check" is. Focus on what the system allowed, not who failed within it.

When This Breaks

When meetings stop happening: Rhythm breaks when people don't see value. Ask: what decision got made in the last three meetings? If the answer is "none," the meetings are performative. Cut frequency or cut the meeting. Some teams work better with async updates. Some need daily check-ins. Find what works for your team and do that instead of following a template that isn't producing decisions.

When transitions between phases are unclear: The Checkpoint questions exist for this. If you can't answer them, you're not ready to transition. That's not a failure; it's the system working. The discomfort of staying in a phase longer than you wanted is less than the cost of moving forward unprepared.

Starting a WISER Implementation

On day one, ensure: everyone knows their Position, the Playbook has initial entries (even sparse), first Witness activities are planned, and team rhythm is scheduled. The Team Rhythm meeting should be on the calendar before the kickoff ends.

Download: After-action review template at wisermethod.com/templates

Note: This is the one Rhythm Play that can be agent-executed. An agent can document the event (from logs and metrics), analyze causes (pattern matching against known failure modes), identify patterns (comparing to past incidents), and draft Playbook updates. The human checkpoint at step 6 ensures accountability for the final action.

Downloads for this chapter at wisermethod.com/templates:

- Team Rhythm agenda template
- Canon Transition Checkpoint template
- Autonomy Graduation Review agenda
- After-action review template

PART V

Part V: Mastery

The Plays work. You've seen them work at Wingman, and by now you've likely tried them yourself. Part V goes beyond Plays to the operational principles underneath them.

These four frameworks separate practitioners who follow WISER from practitioners who own it.

PART V

Chapter 18: Advanced Frameworks

> "The greatest and most robust contribution to knowledge consists in removing what we think is wrong—subtractive epistemology. We know a lot more about what is wrong than what is right." — Nassim Nicholas Taleb, *Antifragile*

Part IV gave you Plays: discrete tools you can pick up and use. This chapter gives you frameworks: mental models that change how you think about the work itself.

Four frameworks. Each one addresses a failure mode we've seen kill implementations that had everything else right.

Risk Burn-Down

Discipline: Address the highest-priority risk first, every time, no exceptions.

A logistics company we worked with had a risk register with forty-seven items. They reviewed it monthly. Every item had an owner and a mitigation plan. By any audit standard, their risk management was excellent.

Their AI routing system failed catastrophically on a Tuesday. Customer shipments went to wrong addresses. The problem wasn't on their risk register. They'd been so busy managing forty-seven risks that they'd never asked: which one could kill us this week?

The Rhythm Plays introduced Risk Burn-down as a 15-minute meeting block. This section teaches the full methodology.

The synthesis: Risk Burn-Down combines Theory of Constraints (address the highest-priority constraint first; everything else waits) with NASA Risk Management (continuous assessment in high-stakes environments where failures compound).

Why this matters for AI: Traditional risk management assumes you have time to notice problems. AI fails faster than humans can recognize patterns. A model that drifts on Monday creates cascading errors by Wednesday. By Friday, you're explaining to leadership why the system approved $200,000 in fraudulent invoices.

At any moment, your team should be able to answer three questions: 1. What's our highest-priority risk right now? 2. What are we doing to reduce it? 3. How do we know if it's working?

If anyone hesitates on any of these, your risk management isn't working.

Scoring Framework:

Factor	Question	Scoring
Probability	How likely is this to occur?	High=3, Medium=2, Low=1
Impact	What's the cost if it happens?	Severe=3, Moderate=2, Minor=1
Detection	Will we know when it happens?	Hard=3, Moderate=2, Easy=1

Priority Score = Probability x Impact x Detection

The detection factor is what most risk frameworks miss. A risk you'll catch immediately is fundamentally different from one that compounds silently. AI systems excel at silent compounding.

Score	Priority	Action
18-27	Critical	Stop other work. Address this now.
8-17	High	Address this week. Active mitigation.
3-7	Medium	Plan mitigation. Monitor.
1-2	Low	Accept or monitor.

The operational discipline:

1. Score all identified risks
2. Address the highest-priority risk first
3. Reduce it until it's no longer highest-priority
4. Move to the next highest
5. Re-score after every mitigation attempt

This sounds obvious. It isn't. Teams naturally want to address easy risks, visible risks, or risks that powerful stakeholders care about. The discipline of "highest score first, no exceptions" prevents political risk management.

Downstream effects:

Before any deployment, map second and third-order consequences. Automation decisions create cascading effects that aren't obvious.

Ask: What dependencies change? What work patterns shift? What information flows get disrupted? What assumptions become invalid?

During Expand, Wingman's Phoenix failure created downstream effects nobody anticipated. Commercial jobs had different billing patterns, and the AI's assumptions about "job complete" didn't hold. The team had focused on first-order risk (will the AI extract data correctly?) and missed second-order risk (what happens when job completion means something different?).

The **Sentinel** owns Risk Burn-Down. It's continuous, not periodic.

Self-Improving Systems

Discipline: Let AI optimize how it works, but never let it redefine what it's working toward.

The rule: AI optimizes how. Humans own what.

This is the line that cannot blur. AI can get better at achieving objectives. It cannot redefine what the objectives are. The moment AI starts adjusting its own goals, you've lost control of the system.

Three stages:

Stage	Description	Human Role
Supervised	Human reviews every adjustment	Approve all changes
Semi-autonomous	AI adjusts within tight boundaries; human audits	Spot-check changes
Autonomous	AI optimizes freely within constraints	Monitor for violations

Most teams want to skip to autonomous. Don't. Each stage builds the trust and evidence required for the next.

What AI can adjust (the "how"):

- Confidence thresholds (within defined ranges)
- Processing efficiency
- Content phrasing
- Routing logic

What AI cannot adjust (the "what"):

- Objectives
- Hierarchy of agency levels
- Boundary constraints
- Escalation triggers

Common failure patterns:

1. **Proxy optimization:** AI improves the metric you gave it while the actual objective declines. You measure response time; AI produces fast but incomplete answers.
2. **Runaway improvement cycles:** Each improvement creates new risks faster than it solves problems. The system is always "getting better" but never stable.
3. **Boundary creep:** AI approaches limits without respecting safety margins. It never violates constraints; it just operates perpetually at 99.9% of them.

The discipline: Every improvement must prove it advances objectives without violating constraints. Baseline before change. Measure after. Compare. If proxy improved but objective didn't, revert.

During Refine, Wingman's team discovered the AI was optimizing extraction accuracy on the data it saw most: HVAC jobs. Plumbing accuracy drifted because the training data was imbalanced. The system was "improving" on aggregate metrics while degrading for a subset of users. Self-improvement without segment-level monitoring is self-deception.

Measurement That Enables Learning

Discipline: Measure what matters, not what's easy to count.

Five criteria:

Criterion	Question	Bad Example	Good Example
Leading	Can you see problems before impact?	Customer satisfaction (lagging)	Error rate by category (leading)
Granular	Guides daily decisions?	Monthly revenue	Conversion by channel by day
Accessible	Both humans and AI can see it?	Locked in reports	Real-time dashboard + API
Gaming-resistant	Hard to hit metric without real outcome?	Ticket closure rate	Closure with customer confirmation
Actionable	Tells you what to do?	"Performance declined"	"Category X errors up 40%"

The proxy trap:

AI optimizes what you measure. Full stop.

If you measure response time when the objective is first-contact resolution, AI produces fast but incomplete answers. If you measure throughput when the objective is accuracy, AI processes more items with more errors.

The solution isn't better proxies. It's measuring both proxy and objective simultaneously. When proxy improves and objective doesn't, stop optimizing the proxy.

Goodhart's Law in practice: "When a measure becomes a target, it ceases to be a good measure." This isn't philosophy. It's operational reality. The moment you tell an AI system to optimize a metric, you've created incentive to game that metric.

Wingman's team measured aggregate extraction accuracy. The AI optimized aggregate accuracy. Nobody measured accuracy by job type until plumbing errors spiked. The metric was accurate; it just wasn't granular enough to catch drift in subpopulations.

The discipline: For every metric you optimize, identify how it could be gamed. Then measure the game. If your metric is "tickets closed," also measure "tickets reopened within 24 hours."

The Rebuild Decision

Discipline: Prove you understand the system before you replace it.

Default: Iterate. Rebuilding from scratch is the extreme exception, not a standard option.

AI makes rebuilds tempting. What used to take months now takes weeks. Teams feel stuck in iteration and think, "Let's just start over." This is almost always wrong.

Why iteration beats rebuild:

Existing systems contain embedded knowledge that nobody documented. Edge cases handled. Integrations tested. Failure modes discovered and patched. A rebuild throws away all of that learning and starts the discovery process over.

The new system won't have the same problems as the old system. It will have different problems, plus new versions of problems the old system already solved.

Before considering a rebuild, all checkpoints must pass:

Checkpoint	Question
Process understanding	Does the team fully understand all critical business logic in the current system?
Technology constraint	Is underlying technology genuinely incapable of iteration?
Manual replication	Can the team replicate the process manually?
Validated objective	Has the objective passed the quality framework?
Rebuild accountability	Is Sponsor willing to accept rebuild risks?

The manual replication test:

Before rebuilding, try running the process manually. If you can replicate it, you understand it well enough to potentially rebuild. If you can't, the existing system contains logic you don't understand, and a rebuild will rediscover that logic the hard way.

When rebuild is justified:

- Manual replication works, proving you understand the process
- Technology is genuinely incapable of iteration (not just difficult)
- Sponsor accepts that rebuild introduces new risks
- Team has capacity to support old system while building new one

Wingman's team never rebuilt. When Phoenix failed, the temptation was to rebuild the extraction model from scratch for commercial jobs. Instead, they iterated: added job status capture, modified the billing trigger logic. The iteration took two weeks. A rebuild would have taken two months and introduced new integration risks.

The discipline: Every rebuild proposal requires explicit answers to all five checkpoints. Fail any one, iterate instead.

These four frameworks share a common thread: discipline over intuition. Risk Burn-Down forces you to address what matters most, not what's easiest. Self-Improving Systems forces you to define boundaries before optimizing within them. Measurement forces you to think about gaming before it happens. The Rebuild Decision forces you to prove understanding before starting over.

WISER isn't a set of ceremonies. It's a set of disciplines. These frameworks are where those disciplines become operational.

Chapter 19: Creating Your Own Plays

"If you can't describe what you are doing as a process, you don't know what you're doing." — W. Edwards Deming

The Starter Plays Are a Starting Point

The Plays in Part IV are not mandatory. They're general-purpose tools that work across industries and team sizes. Use them as-is, modify them, or replace them entirely.

What matters is that certain functions get served. You need alignment (Playbook or equivalent). You need discovery (some form of mapping). You need governance (some form of oversight). The specific Play is flexible. The function is not.

The rule: Principles are non-negotiable. Canons guide the sequence but allow iteration. Plays are yours to adapt.

Play Types

Every Play in the Starter Plays falls into one of seven categories:

Type	Purpose	Examples
Foundational	Team structure and roles	Seven Positions, DACI
Mapping	Discovery and documentation	Friction Map, User Flow Map
Testing	Assumption validation	Experiment Types, Assumption Inventory
Building	Implementation guidance	Objective Quality Framework, Pilot Planning
Scaling	Expansion patterns	Context Comparison, Rollout Checklist
Governance	Oversight and control	Hierarchy of Agency, Drift Monitoring
Rhythm	Cadence and checkpoints	Team Rhythm, Canon Transitions

When you create new Plays, they'll fit into one of these categories. This helps teams know where the Play plugs in.

Modifying Starter Plays

Most teams don't need to create Plays from scratch. They modify what's here.

Common modifications:

- Adding fields to templates (industry-specific compliance requirements)
- Removing sections that don't apply (small teams skip DACI entirely)
- Changing terminology (your organization calls Positions something else)
- Adjusting thresholds (different confidence levels for Graduation decisions)

The test: Does your modification preserve the Play's intent? A Friction Map that doesn't capture where work breaks down isn't a Friction Map variant; it's a different tool.

Replacing Starter Plays

Sometimes your organization already has something better.

If your company has a proven stakeholder mapping methodology, use it. If your team has a review format that consistently produces insights, keep it. The Starter Plays aren't trying to replace what works.

When to replace:

- Your existing tool serves the same purpose
- Your team already knows how to use it
- It produces reliable results

When not to replace:

- Your existing tool is "how we've always done it" without evidence it works
- You're avoiding learning something new
- The replacement doesn't actually serve the Play's purpose

Creating Your Own Plays

Plays come from two paths: discovery and design.

Discovery: You notice you're repeating something. After enough implementations, the pattern becomes clear enough to codify. You're recognizing a Play that already exists in your practice.

Design: You know what needs to happen. You break it into executable tactics upfront. You're creating a Play to accomplish a known objective.

Both paths produce valid Plays. The difference is whether you're capturing what you learned or planning what you'll do.

The Discovery Path

Sometimes you don't know a pattern exists until you've lived it.

The three-implementation rule:

- After one implementation: you have anecdotes
- After two implementations: you have hypotheses

- After three implementations: you have a pattern worth codifying

If you're explaining the same thing for the third time, it's a Play. Write it down.

When to use the discovery path: You're doing work you haven't done before. You don't know what the pattern is yet. Let experience reveal it.

The Design Path

Sometimes you know exactly what needs to happen. You don't need to fail three times first.

You have an objective. You understand the domain. You break the work into executable tactics. Each tactic becomes a Play.

When to use the design path: You're planning work where the steps are knowable in advance. You're decomposing a Playbook into its component Plays. You're engineering a solution, not discovering one.

The three-implementation rule doesn't apply here. You're not waiting to recognize a pattern. You're creating the pattern deliberately.

The risk: Designing Plays for work you don't actually understand. If you're guessing at the steps, you're better off on the discovery path. Design works when you have genuine clarity about what needs to happen.

Play Structure

Every Play follows the same structure. This makes Plays predictable and usable without reading instructions.

Section	What It Answers
Context	When should you use this Play?
Objective	What will it accomplish?
Inputs	What do you need before starting?
Position	Which Position owns or executes this?
WISER Fit	Where in W-I-S-E-R does it apply?
Steps	What do you do, in order?
Tools	What templates or resources are needed?
Pitfalls	What commonly goes wrong, and what do you do about it?
Variations	How do you adapt for different conditions?
Success	How do you know it worked?

A Play is a reusable mini-plan. Whoever runs it can apply it with confidence without reinventing the wheel.

Writing Each Section Well

The difference between a useful Play and a useless one is precision. A Play that tries to cover every edge case becomes unusable. Document the core path. Note the pitfalls. Keep it lean.

Context: State what triggers the Play and what disqualifies it. "Use when X" is incomplete. "Use when X, not when Y" draws a boundary.

Objective: Name the outcome and how you'll verify it. Vague: "Improve routing." Precise: "Route cases to the right path within SLA." The precise version tells you what to measure.

Inputs: List what you need before starting, with enough specificity to know when you have it. "Data" is useless. "Volume by category, last 90 days" is actionable.

Position: Name who owns the outcome and who executes. If multiple Positions are involved, clarify handoffs.

WISER Fit: State which Canon this Play primarily supports. Some Plays span multiple Canons; note that.

Steps: Include decision points, not just actions. "Set threshold" is incomplete. "Set threshold; start conservative, tighten as confidence grows" tells you how to navigate the choice. If a step requires judgment, say what informs that judgment.

Tools: List templates, worksheets, or resources the Play requires. If none, say "None."

Pitfalls: Name the failure mode, then name the response. "Threshold too loose" is a warning. "Threshold too loose; start at 20% human review and adjust based on miss rate" is guidance.

Variations: Note how the Play adapts for different conditions. If a variation changes the steps significantly, it's probably a separate Play.

Success: Quantify when possible. "Works well" is subjective. "90%+ within SLA, <5% wrong calls" is verifiable. If you can't quantify, describe what observable evidence would confirm success.

What Makes a Good Play

A Play is good if whoever runs it can do so without you in the room.

Clarity test: Can someone unfamiliar with your context follow it?

Completeness test: Does it produce the output it promises?

Elegance test: Can you remove anything without degrading execution? If yes, cut it.

Principle test: Does it conflict with any AI First Principle? If your Play violates a principle, it's not a WISER Play. Review the principles in Chapter 4 if you're unsure.

Novelty test: Does it add something the Starter Plays don't cover, or does it merely rebrand existing Plays?

A good Play disappears. People stop noticing it's there. They just use it, get the result, and move on. That's the goal.

Downloads: Play creation template at wisermethod.com/templates

Chapter 20: When WISER Fails

> "Knowing what you cannot do is more important than knowing what you can do." — Lucille Ball

WISER is powerful. It's not universal.

Every method has boundaries. Knowing where those boundaries are prevents misapplication and builds trust with the people who will use what you build. This chapter names the conditions where WISER fails or shouldn't be applied.

No Senior Leadership Buy-In

This is the most common failure and the most fatal.

WISER requires permission. Permission to slow down during Witness instead of jumping to solutions. Permission to run experiments that might fail. Permission to expand incrementally instead of rolling out everywhere at once. Permission to tell the truth when timelines slip.

Without executive air cover, the first setback kills the initiative. A sponsor who can't protect the process can't protect the team. And a team without protection will abandon the method the moment pressure arrives.

The signs are clear: leadership says "yes" to WISER but assigns no budget. They approve the pilot but demand results in two weeks. They want transformation but won't attend the steering meetings. They nod at observation and then ask why you haven't started building yet.

If your sponsor can't protect the process, the process cannot protect you.

What to do: Before starting, have an explicit conversation with your sponsor about what protection looks like. What happens when the first delay hits? Who handles the executive who wants to skip Witness? If the sponsor can't answer these questions with confidence, you don't have buy-in. You have polite tolerance. That's not enough.

Speed Genuinely Matters More Than Learning

Sometimes you don't have time to observe. Crisis response. Market windows. Regulatory deadlines. Competitive threats that will close the opportunity before you can study it.

WISER's strength is learning before building. But learning takes time. If the building is on fire, you don't stop to understand the building.

The key word is "genuinely." Most urgent projects are not actually urgent. Leadership declares urgency because waiting is uncomfortable, not because waiting is impossible. The team that shipped fast and shipped wrong will spend longer fixing the mistakes than the team that took two extra weeks to observe.

But sometimes urgency is real. If your competitor will own the market in ninety days and your observation phase takes sixty, you may need a different approach. If the regulator requires compliance by a fixed date and discovery would push you past it, compliance comes first.

What to do: When someone declares urgency, ask: "What happens if we take two more weeks?" If the answer is "nothing, really," the urgency is manufactured. If the answer involves concrete consequences, competitive losses, or regulatory penalties, the urgency may be real. Proceed accordingly.

Culture Punishes Experimentation

WISER requires psychological safety. The method assumes you can surface problems without being blamed for them. You can run experiments that fail without being punished. You can tell executives uncomfortable truths without ending your career.

Some organizations punish all of this.

If surfacing problems gets people fired, problems stay hidden. If failed experiments become performance review fodder, nobody runs experiments. If the messenger gets shot, messages stop arriving.

The method cannot fix a culture that fears transparency. You can try to create a protected bubble where the WISER team operates by different rules, but bubbles pop. Eventually the organizational antibodies find you.

What to do: Observe how the organization handles failure before proposing WISER. Ask about the last project that didn't work. What happened to the team? Were they blamed or celebrated for learning? If the answer involves scapegoating, reorganizations, or career consequences, the culture isn't ready. You might still try a small pilot with a very protected sponsor, but go in with eyes open.

Overhead Exceeds Complexity

WISER has overhead. Playbooks. Positions. Rituals. Documentation. For complex, uncertain problems with significant organizational stakes, this overhead pays for itself. For simple problems, it doesn't.

A small team automating a single report doesn't need a Playbook. A two-week project with a clear solution doesn't need formal Witness. A well-understood workflow with obvious friction points doesn't need extensive mapping.

The method should serve the work, not the reverse. Using WISER to feel rigorous when simpler approaches would work is a waste of everyone's time.

What to do: Before applying WISER, ask: "What's the cost of getting this wrong?" If the answer is "we fix it next week," you probably don't need the full method. If the answer involves organizational disruption, significant investment, or changes that are hard to reverse, the overhead is justified. Scale the method to the stakes.

The Problem Is Deterministic

WISER is designed for uncertainty. Problems where the right answer isn't known in advance. Problems where context matters. Problems where the system needs to learn and adapt.

Some problems don't have that kind of uncertainty. They have rules.

If every input maps to exactly one correct output, and you can write those rules down, you don't need AI and you don't need WISER. A conditional statement handles it. A lookup table solves it. The friction isn't uncertainty; it's just implementation.

But be careful. The most common failure isn't applying WISER to deterministic problems. It's assuming a problem is deterministic when it isn't. Leaders often believe they understand a system completely until observation reveals edge cases nobody anticipated. The dispatcher's spreadsheet looked like a workaround until watching the actual work revealed it was the only thing that worked.

The signs of a genuinely deterministic problem: the process has been stable for years *and* you've recently observed it. The edge cases are known, finite, and documented. Multiple subject matter experts independently produce the same decision tree. When you ask "what should happen if X?" the answer is always the same *and* you've tested that claim against real data.

What to do: Before concluding a problem is deterministic, spend at least a few days in Witness mode. Watch the actual work. If your decision tree survives contact with reality, proceed without WISER. If observation reveals cases the tree doesn't handle, the problem isn't as deterministic as you thought.

High-Stakes Environments Where Failure Isn't Safe

WISER's approach to learning involves controlled failure. Small experiments. Pilots that might not work. Incremental expansion that tests assumptions. This requires an environment where small failures don't cascade into catastrophes.

Some environments can't tolerate that.

Healthcare systems where a bad prediction could harm patients. Aviation systems where errors compound faster than humans can intervene. Financial systems where a small mistake triggers regulatory consequences. Nuclear facilities. Critical infrastructure. Anywhere that "fail small and learn" could mean "fail once and cause irreversible harm."

WISER's experimental approach may not be safe in these contexts. You might use the observation and mapping disciplines while applying more rigorous validation before any deployment. Or you might need a different method entirely.

What to do: Identify your failure tolerance early. Can you run experiments in production? Can you pilot with real users? If the answer is "not safely," you need either a simulation environment robust enough to represent reality or a different methodology designed for high-stakes deployment. WISER can inform discovery; it may not be appropriate for deployment.

Naming these limits isn't weakness. It's the same discipline WISER asks you to apply to AI systems: know what you're working with, understand where it breaks, and don't pretend otherwise.

Some practitioners read this chapter and feel deflated. They wanted WISER to be the answer to everything. It isn't. No method is.

Other practitioners read this chapter and feel relief. They've been in organizations with no executive buy-in, cultures that punish failure, or high-stakes environments where experimentation wasn't safe. They wondered if they were doing something wrong. They weren't. They were applying a method to contexts where it couldn't succeed.

Choosing the right tool for the context is the point. If WISER isn't right for your situation, honor that. Find what is. The discipline of honest assessment you learned here transfers to whatever approach you use next.

PART V

Chapter 21: Integrating Methods

> "The test of a first-rate intelligence is the ability to hold two opposing ideas in mind at the same time and still retain the ability to function."
> — F. Scott Fitzgerald

Most practitioners don't start from zero.

You arrive with Lean. Or Agile. Or Six Sigma. Or Design Thinking, EOS, ITIL, SAFe, BPM, or some hybrid your organization assembled over years of trying different approaches. You have rituals, terminology, and muscle memory built around those methods.

You don't have to throw them out. You reframe them.

WISER becomes the lens through which you view your existing practices. The rituals you know become tools within WISER's framework. Your standups, your value stream maps, your control charts; they still have value. But now they serve a different purpose: supporting the questioning discipline that WISER requires.

What WISER Adds

Most methodologies share a gap. They assume you know what to build before you start building.

Lean maps the value stream, then optimizes it. But it rarely asks whether the process should exist at all. Agile fills the backlog, then executes it. But where did that backlog come from? Six Sigma measures variation, then reduces it. But what if the entire workflow should be replaced? Design Thinking interviews users, then ideates solutions. But a week of interviews isn't the same as watching how work actually happens.

Each methodology excels at execution. WISER excels at questioning.

The discipline WISER adds:

- **Observation before action.** Don't trust documentation. Watch what people actually do.
- **Question existence, not just efficiency.** Before optimizing a process, ask whether it should exist. This is a Lean discipline: eliminate before you simplify, simplify before you automate. The most common engineering mistake is optimizing something that shouldn't exist.
- **Test assumptions before scaling.** Experiments reveal what planning conceals.
- **Accept that AI changes the territory.** Methods designed for deterministic software need adaptation for probabilistic systems. Six Sigma was designed for stable processes. AI systems learn and change. A process in control yesterday might drift out of control tomorrow because the model updated.

Your methodology gives you rigor. WISER gives you the habit of asking whether you're applying that rigor to the right problem.

The Pattern Across Methodologies

Every methodology risks the same failure: optimizing the wrong thing with precision.

Methodology	Core Strength	The Gap
Lean	Eliminates waste in known processes	Assumes you know what waste looks like
Agile	Iterates quickly on a backlog	Assumes the backlog addresses the real problem
Six Sigma	Reduces variation through measurement	Assumes the process should exist and stay stable
Design Thinking	Centers user needs in ideation	Assumes interviews reveal what observation would
EOS	Creates accountability through 90-day cycles	Assumes you can scope the work before observing it
ITIL	Standardizes service management	Assumes standards match current reality
SAFe	Coordinates Agile at scale	Assumes coordination doesn't become ceremony
BPM	Documents and improves workflows	Assumes documentation reflects what people do

The table reveals a consistent pattern. Each methodology has a core strength that WISER doesn't replace. Lean excels at eliminating waste once you know what waste looks like. Agile delivers working software in short cycles once you know what to build. Six Sigma reduces variation once you've identified the right process to control. Design Thinking generates creative solutions once you understand the problem space.

Look at the gap column. Every methodology assumes discovery ends. The gap is the same across all of them: the assumption that discovery is a phase you complete, not a discipline you maintain.

WISER's contribution isn't another execution framework. It's the questioning layer that sits beneath whatever framework you use. Before you map the value stream, watch what actually happens. Before you fill the backlog, test whether those items should exist. Before you measure variation, ask whether the process itself should be replaced.

Consider Agile's Sprint 0. Many teams acknowledge the need for discovery before building, so they add a sprint for research and design. But Sprint 0 is often treated as an awkward prerequisite, something to get through quickly so the team can start delivering points. Observation gets compressed into a week or two, then shelved once velocity becomes the metric that matters. WISER puts observation at the center of the method, not as a warm-up act.

The Common Failure Mode

All methodologies can fail the same way: when they become ceremony.

Standups that no one listens to. Retrospectives that change nothing. Value stream maps that sit in shared drives unread. Control charts that nobody checks. Process documentation that describes a reality from 2019.

The ritual happens. The meeting occurs. The form gets filled out. But nothing changes. The methodology becomes its own purpose rather than serving the work.

WISER's first job is to surface that gap. Watch what people actually do. Compare it to what the methodology says they should do. Name the difference.

That difference is often where the value lives.

Reframing in Practice: A Team's Story

A financial services company ran Scrum across twelve development teams. They had the rituals down: sprint planning, daily standups, retrospectives, quarterly PI planning. Velocity was tracked. Burndown charts were reviewed. The methodology was mature.

They brought in an AI initiative to automate document processing. The product owner wrote user stories. The team estimated in points. Sprint 1 began.

By Sprint 3, they had built a document classifier that worked in demos but failed in production. The training data came from a sample the product owner had curated. Real documents looked different. Edge cases multiplied. The team kept sprinting, but they were sprinting in the wrong direction.

When they reframed their process through WISER, the first change was small. Before Sprint 4 planning, two team members spent three days in the operations center. Not interviewing. Watching. They saw clerks manually routing documents the classifier couldn't handle. They saw workarounds nobody had documented. They saw the actual problem, which wasn't classification accuracy. It was the 40% of documents that didn't fit any category.

That observation changed the backlog. Instead of improving the classifier, they built a triage system that routed ambiguous documents to human review with AI-suggested categories. The clerks became editors instead of processors. Accuracy improved because the humans and AI worked together instead of the AI trying to replace humans entirely.

The team kept running Scrum. Same ceremonies. Same cadence. But the ceremonies now served a different purpose. Sprint planning started with a Witness checkpoint: someone had to report what they'd observed in the past sprint. Not metrics. Observations. What surprised them. What users actually did with the system. What the AI got wrong in ways the dashboards didn't capture.

The Scrum Master pushed back. "We already observe," she said. "That's what sprint reviews are for. We demo to stakeholders every two weeks."

The response: "Sprint reviews show stakeholders what you built. They don't show you what users do with it. Reviewing your own output isn't the same as watching someone use the system in their environment, with their workarounds, under their time pressure. When was the last time someone on this team sat in the operations center for a full shift?"

That distinction landed. The team had been demoing to stakeholders who nodded approvingly. They hadn't watched the clerks who actually processed documents curse at the system and route 40% of cases manually.

Reframing, Not Replacing

The goal isn't to run two methodologies in parallel. That's overhead.

You keep the container. WISER changes what goes inside it.

Reframing Your Existing Practices

You don't need all of the reframings below. Find the methodology you use and start there. Each section includes a reframing table and guidance on what AI changes for your specific practice.

Agile/Scrum

Existing Ritual	Reframed Through WISER
Sprint Planning	Add "What did we observe?" as the first agenda item.
Mid-Sprint Check-in	Add a dedicated 30-minute mid-sprint session for observation reports.
Sprint Review	Include one observation from production — a story, not a metric.
Retrospective	Add: "What assumption did we hold that turned out to be wrong?"
Backlog Refinement	For each item, ask: "What evidence do we have that this should exist?"

What AI changes for Agile/Scrum:

On estimation: AI work breaks traditional estimation. A story that looks like three points becomes thirteen when the model behaves unexpectedly. Instead of estimating the unknowable, reserve capacity for discovery. Treat 15-20% of sprint capacity as observation and experimentation time that doesn't get story-pointed. You're not slowing down; you're acknowledging that some work can't be sized until you've learned what you're actually building.

On the Product Owner relationship: "Question whether this item should exist" can feel like a challenge to PO authority. It isn't. The Scout role in WISER is a questioning function, not a backlog veto. The PO still owns prioritization. What changes is that backlog items now require evidence, not just stakeholder requests. The Scout surfaces observations; the PO decides what to do with them. Frame it as giving the PO better information, not overriding their judgment.

See also: Handling Resistance and The Questions That Matter below — these apply across all methodologies.

Lean/Value Stream

Existing Practice	Reframed Through WISER
Gemba Walks	Watch for new patterns: silent failures, drift that looks like normal variation, outputs changing without process change.
Waste Identification	Map the hidden factory: human workarounds that compensate for AI limitations.
Kaizen Events	Dedicate the first half-day to observing the current process with AI behaviors included.
Standard Work	See below — AI fundamentally challenges Standard Work's assumptions.

What AI changes for Lean/Value Stream:

On Standard Work: Standard Work assumes repeatability. AI doesn't offer it. The same input can produce different outputs; the model can drift between shifts. This breaks the fundamental premise. The adaptation: Standard Work for AI-assisted processes defines the human decision points and acceptable output ranges rather than scripting the full process. Document when a human reviews AI output, what triggers escalation, and what "good enough" looks like. The AI's behavior becomes a variable the Standard Work accommodates, not a step it scripts.

On scope: AI disrupts foundational Lean concepts like Takt time and Statistical Process Control. Adapting these tools for probabilistic systems is beyond this chapter's scope, but the observation discipline applies: watch how the AI actually behaves before assuming your existing control mechanisms still work. The same applies to supplier and customer interfaces in the value stream — when AI changes what you deliver or what you need, that's a mapping exercise this chapter doesn't cover.

See also: Handling Resistance and The Questions That Matter below — these apply across all methodologies.

Six Sigma

Existing Practice	Reframed Through WISER
Define Phase	Shift the question: "How can AI assist this required process while maintaining compliance?"
Measure Phase	Include observational data alongside metrics — with operational definitions so observation counts as data.
Analyze Phase	Test assumptions through experiments within your existing validation protocols.
Control Phase	Monitor for model drift, data drift, and concept drift — traditional control charts assume stable variation AI doesn't provide.

What AI changes for Six Sigma:

On scope: Six Sigma practitioners in regulated industries will note gaps: Measurement System Analysis (MSA) for AI classifiers, Failure Mode and Effects Analysis (FMEA) adaptation for probabilistic failure modes, control chart methodology for non-stationary processes, human-AI handoff documentation. These are real challenges deserving dedicated treatment beyond this chapter's integration lens. For regulatory constraints more broadly, see Chapter 20's discussion of where WISER doesn't apply.

See also: Handling Resistance and The Questions That Matter below — these apply across all methodologies.

Design Thinking

Existing Practice	Reframed Through WISER
Empathize	Watch for AI-specific patterns: unreported workarounds, inconsistent behavior, interactions that change when the model updates.
Define	When the problem is mandated, shift to: "Is this the right framing of what we've been asked to solve?"
Ideate	Include "this interaction shouldn't exist" as a valid design direction.
Prototype	Use Wizard of Oz testing to make AI behavior tangible before the model exists. See Chapter 13.
Test	Continue observing after launch — AI reveals drift and compounding edge cases, not just usability issues.

What AI changes for Design Thinking:

On AI as design material: Traditional Design Thinking treats technology as implementation — you design the experience, engineers build it. AI is different. It has behaviors, limitations, and surprises that shape the design space. You're not designing for a tool. You're designing with a collaborator that has its own tendencies. This changes prototyping, testing, and iteration — the AI's behavior is part of the design material, not just the delivery mechanism.

On scope: AI explainability — helping users understand what the system is doing and why — affects every phase of design. It's a challenge that deserves its own treatment. For cross-functional collaboration between designers and ML engineers, see the Positions framework in Chapter 10.

See also: Handling Resistance and The Questions That Matter below — these apply across all methodologies.

EOS (Entrepreneurial Operating System)

Existing Practice	Reframed Through WISER
L10 Meeting (Headlines)	Use Headlines to surface one observation from the field: something someone saw that the Scorecard didn't capture.
Rocks	Before setting an AI Rock, ask: what have we observed that tells us this is the right problem to commit to for ninety days?
Scorecard	Add a measurable for AI system health: human override rate, extraction accuracy, or AI-assisted decision volume.
Issues List (IDS)	When AI issues surface, the Solve is often an observation action: assign someone to watch the process and report findings at the next L10.
Accountability Chart	AI implementation crosses seat boundaries. Name who owns the AI Rock: the person who owns the process being changed, not a centralized AI function.

What AI changes for EOS:

On Rocks: EOS runs on 90-day cycles. Set a Rock, commit, execute, report. That cadence works for initiatives you can scope. AI work often can't be scoped that way. A team sets a Rock — "Deploy AI-assisted claims processing" — and by week six, they've learned the real problem isn't claims processing at all. The Rock was set before anyone observed the work. The fix isn't abandoning quarterly Rocks. It's changing what the first Rock looks like. Make the first AI Rock an observation Rock with a concrete deliverable: "Deliver a documented assessment of the three highest-value AI opportunities in claims with estimated impact." That's achievable. It prevents the next three quarters of Rocks from solving the wrong problem.

On the Visionary/Integrator dynamic: AI initiatives tend to originate with the Visionary. They see the potential, they want to move. The Integrator's job is to make it operational. WISER sits in the tension between them. The Visionary's instinct is to skip observation and start building. The Integrator's instinct is to plan before acting. WISER says both are wrong until someone has watched the work. The Visionary needs to hear that observation accelerates the vision rather than delaying it. The Integrator needs to hear that observation replaces planning, not adds to it.

On seat ownership: The deeper problem is where AI lives on the Accountability Chart. If the AI Rock lives with the Visionary, it never gets operational. If it lives with the Integrator, the Visionary feels sidelined. If it lives with a department head, it becomes siloed. The answer: AI Rocks belong to whoever owns the process being changed. Claims processing AI is an Operations Rock. Sales automation is a Sales Rock. The AI capability is a tool, not a seat. This prevents AI from becoming an orphan initiative that everyone sponsors and nobody owns.

On scope: Deeper EOS integration — where AI lives in the V/TO, how it changes the People Analyzer, multi-level L10 coordination — deserves its own treatment beyond this chapter.

See also: Handling Resistance and The Questions That Matter below — these apply across all methodologies.

Handling Resistance

Reframing meets friction. Here's what you'll hear and how to respond.

"We already do discovery."

Ask when they last observed users in production. Not interviewed. Watched. Most teams confuse requirements gathering with observation. Requirements capture what stakeholders say they want. Observation reveals what actually happens.

"This will slow us down."

Observation prevents building the wrong thing. Building the wrong thing fast is not velocity. It's waste with good metrics. A three-day observation that redirects a six-week sprint isn't slowing down. It's preventing six weeks of wasted effort.

"Our methodology already handles this."

Ask which ritual surfaces the gap between what the process document says and what people actually do. Most methodologies have no mechanism for this. They assume the documentation reflects reality.

"Leadership wants to see velocity, not observation reports."

Frame observation as risk reduction. Every assumption you test before building is a failure you avoid after deployment. Leadership cares about outcomes. Show how observation improves outcomes, and the velocity conversation shifts.

"The AI team handles AI. We handle the process."

AI implementation is a process change disguised as a technology project. The teams that succeed are the ones where process people and AI people observe together. The process person sees what work looks like. The AI person sees what's technically possible. Neither perspective alone is sufficient.

The Questions That Matter

WISER adds specific questions to whatever methodology you use. These questions don't replace your existing rituals. They sharpen them.

Before you build:

- What did we observe that we didn't expect?
- What assumption are we making that we haven't tested?
- Should this feature exist at all, or are we optimizing something that should be deleted?

While you build:

- What is the AI doing that surprises us?
- Where are humans working around the system instead of with it?
- What edge cases are we discovering that weren't in the requirements?

After you deploy:

- How does actual usage differ from expected usage?
- What workarounds have emerged?
- Where has the AI drifted from its original behavior?

These questions fit into any existing meeting structure. They don't require new rituals. They require a commitment to questioning assumptions rather than defending them.

For multi-team coordination and SAFe environments, the reframing pattern is the same; the coordination complexity is another matter entirely.

If you can't change the rituals yet, you can still change what you pay attention to. Bring one observation to the next meeting. Ask one assumption question. Document one surprise. The reframing starts with attention, not authority.

Why This Works

Your existing toolkit isn't wrong. It needs a new lens.

WISER provides that lens. The result is the same whether you use it on its own or through practices you already know: the discipline to question what you're building before you build it.

PART V

Chapter 22: The Journey Continues

"The only sustainable competitive advantage is an organization's ability to learn faster than its competition." — Peter Senge, *The Fifth Discipline*

The Merge

Chapter 1 told you about the children in the Colombian jungle. Thirty-five days of precision found nothing. One night of vision found everything. Neither approach alone would have found them.

WISER teaches that merge. Discipline plus empathy. Rigor plus observation. The project management muscle you already have, combined with a way of thinking that treats uncertainty as the territory, not a problem to solve before entering it.

This book gave you both approaches. Now you have to use them.

What Mastery Looks Like

Mastery isn't a destination. It's pattern recognition you can only build through practice.

After enough implementations, you'll notice:

You see the friction before they describe it. Five minutes into observation, you know where the work breaks down. Not because you're smarter. Because you've seen the patterns enough times to recognize the shape.

You smell untested assumptions. When someone says "users will love this" or "the data is clean enough," something in your gut fires. You've heard those words before. You know what comes next.

You know when to iterate back. The Canon boundaries become less rigid in your head. You're not following a checklist. You're sensing when the implementation needs more observation, more testing, more validation. The method becomes intuition with a structure to hold it accountable.

You create Plays without thinking about it. That adaptation you made for healthcare compliance? That's a Play. The way you handle multi-stakeholder alignment in government? That's a Play. You're not just using the Starter Plays. You're extending the method.

You hold the process under pressure. When the executive says "just ship it," you can explain why that destroys value without being preachy about it. When the team wants to skip Witness, you know how to make the case in terms they'll hear. The method gives you language for conversations that used to be impossible.

Mastery shows in what you don't do. You don't skip observation because you're excited to build. You don't expand before you're ready. You don't grant autonomy without evidence. You don't ignore drift because the metrics look fine.

The discipline becomes invisible. But it's always there.

The Moment It Clicks

Robb tells this story about a financial services engagement.

He walked into a kickoff meeting. The CTO wanted AI-powered fraud detection. The CFO wanted AI-powered cost reduction. They were both nodding. They both said "alignment." They both thought they were talking about the same project.

They weren't.

Robb had seen this before. Two executives, same meeting, different movies playing in their heads. The CTO's fraud detection would require more analysts, not fewer. The CFO's cost reduction meant headcount. Neither had said it out loud. Both assumed the other understood.

He asked one question: "When this succeeds, what changes for your team?"

The CTO said "We catch fraud we're currently missing."

The CFO said "We reduce manual review by 40%."

The room got quiet.

Without that question, they would have built for six months before discovering they had incompatible definitions of success. Robb didn't need to run the project to know this. He'd seen the pattern. When two sponsors nod at the same words but describe different outcomes, the project is already dead. You just don't know it yet.

That's mastery. Not brilliance. Pattern recognition.

You'll have your own version of this moment. You'll be in a room, and something will feel off. You won't be able to explain it immediately. But you'll know to ask the question that nobody else is asking. And you'll be right.

The method teaches you what to look for. Practice teaches you to see it without looking.

The Real Competition

Models are commoditizing. Implementation capability isn't.

Every major tech company offers foundation models. Open-source alternatives multiply monthly. The technology itself is becoming table stakes. In five years, access to powerful AI won't differentiate anyone.

What will differentiate: the ability to implement. To observe operations clearly. To test assumptions honestly. To build systems that actually work in production. To expand without breaking. To govern without strangling.

The companies winning at AI aren't the ones with the best models. They're the ones who can implement, learn, and iterate faster than their competition.

This book gave you the method. Whether you build the capability is up to you.

The Question That Changes Everything

The scarcity mindset asks: "How do we do the same work with fewer people?"

The abundance mindset asks: "What work becomes possible that wasn't before?"

Wingman's billing team didn't shrink. It grew into something new. Clerks who used to chase data entry errors became exception handlers, catching what the AI missed. The work wasn't eliminated. It was elevated. That's the difference between automation and transformation.

The method you just learned is designed for the second question. It assumes AI creates capability, not just efficiency. It assumes the goal is better outcomes, not smaller headcounts. It assumes the people closest to the work have insight worth preserving.

Not every organization makes this choice. Some will use WISER to cut costs and call it transformation. They'll get what they're looking for. But they'll miss what they could have built.

The question you answer at the start shapes everything that follows.

Your First 90 Days

You don't need permission to start observing. You don't need a team to start asking questions. The method scales down to a single person with a notebook. (Chapter 20's warning about leadership buy-in applies to organizational deployment. Individual observation is how you build the case for that buy-in.)

Days 1-30: Orient

Read the Starter Plays in Part IV even if you're not using them yet. The structure will feel abstract until you see a problem it applies to. Identify one process in your organization that smells wrong. Something people complain about. Something that should be faster or simpler than it is. Start observing it informally. You're already doing Witness.

Find one potential ally. Someone who sees the same friction you see. Someone who might become a Sponsor or a team member if you can show early results.

Days 31-60: Experiment

Propose a small pilot. One problem, one context, two to four weeks. Use the Starter Plays or adapt them. The templates matter less than the discipline: observe before building, test before scaling, document what you learn.

Create a Playbook even if it feels premature. Run one Friction Map session with whoever will participate. The quality of the output matters less than the quality of the conversation.

Days 61-90: Prove

Complete one Witness-Interrogate-Solve cycle. It doesn't have to be elegant. It has to produce a result you can point to.

Capture the results, even if modest. Present to stakeholders: what you observed, what you tested, what you built, what happened. Then decide: expand, refine, or pivot to a different problem.

Start small. Learn what works. Then make the case for more.

You're Not Alone

Practitioners around the world are running WISER implementations right now. They're adapting Plays for healthcare, government, manufacturing, finance. They're discovering what works in contexts we haven't seen. Some started exactly where you are: one person, one notebook, one process that smelled wrong.

The method is living documentation. It improves through practice. When you adapt a Play or discover something that breaks, that learning feeds back. You're not just using WISER. You're extending it.

Everything you need to start is in this book. For structured practice with experienced practitioners, facilitated cohorts, and a community working through the same challenges, visit wisermethod.com.

One Last Thing

The children in the jungle weren't found by better grids. They were found when someone stopped searching and started seeing.

Your organization's value is still out there. You now have a different way to search.

Go find it.

PART V

Appendix A: Parking Lot

Ideas Worth Preserving

Some concepts didn't fit the core method. Not because they're wrong, but because they add complexity without clear benefit for most practitioners, or because they're questions we haven't resolved yet. We preserve them here.

If implementations reveal patterns we haven't seen, some of these ideas may earn their place in future versions. For now, they wait.

Objective-Oriented Programming

Object-oriented programming organizes code around modular objects with defined behaviors. "Objective-oriented" thinking could organize work around modular objectives with defined measurements and constraints. The parallel captures something real about how WISER works: objectives become first-class citizens, clear boundaries enable autonomous operation, and hierarchical structure manages complexity.

The programming metaphor may alienate non-technical audiences, and the method works without it. We're also uncertain whether forcing the parallel adds clarity or confusion.

Possible future use: Advanced training for technical audiences. Tooling development. Academic positioning.

Nested Objectives and Hierarchy of Agency

How do nested objectives (goal decomposition) relate to hierarchy of agency (decision authority)? Both use tree-like structures, but for different purposes.

Our current read: these are orthogonal concerns. A project could have a single objective with complex agency hierarchy, or deeply nested objectives with simple agency structure. The relationship may correlate in practice but isn't necessarily structural. We don't want to force a connection prematurely.

Questions we're tracking:

- Do objective characteristics predict appropriate agency levels?
- Does objective complexity correlate with agency hierarchy complexity?
- When teams revise objectives, does hierarchy of agency change?

Monozukuri

A Japanese concept meaning "making things." But the translation misses its essence. Monozukuri is the philosophy that making is itself a craft. You cannot separate the quality of what you make from how you make it, or from who is making it.

WISER's emphasis on learning through iteration, discovery before disruption, and building from user experience all echo these principles. The philosophy validates why you can't skip Witness, why Interrogate must precede Solve, why the people doing the work must be part of the redesign. But those principles are already embedded in WISER without requiring foreign terminology. Adding the Japanese term risks appearing pretentious without adding practical value.

Possible future use: A one-sentence lineage acknowledgment in Chapter 21's Lean section or the Author's Note, honoring the connection without adopting the terminology. Academic contexts where intellectual lineage matters.

The Simplification Sequence

Question the requirements. Eliminate unnecessary parts. Optimize what remains. Scale and economize. Each step must complete before the next begins.

The sequence directly challenges premature optimization. It forces teams to ask whether components should exist before asking how to make them efficient. But the Canons already handle this: Witness (observe and question), Interrogate (test assumptions), Solve (build what survives). Adding another sequence would create competing frameworks.

Possible future use: Marketing and communication contexts where a crisp compression of WISER's logic is needed: back-cover copy, conference talks, elevator pitches. The sequence isn't a competing framework; it's a distillation of the existing one. Also domain-specific Plays for manufacturing, engineering, or process optimization.

Complexity Theory Grounding

Complex systems that work in nature are typically outputs of simple systems iterated repeatedly. The reverse, designing complex systems upfront, typically fails. The embedded complexity can't be managed.

This provides theoretical grounding for WISER's iteration-first approach. AI breakthroughs follow the same pattern: simple algorithms plus massive iteration equals sophisticated results. But the book demonstrates this through practice; adding theory would break the cognitive flow. The case study shows emergent complexity from simple iterations. Readers internalize the principle without academic framing.

Possible future use: Academic papers. Materials for audiences who value theoretical foundations.

Optimal Stopping

When do you stop iterating? Principle 12 says justify resource consumption. The canon checkpoints in Chapter 17 create evaluation points. But neither provides explicit guidance for the judgment call: is this good enough, or should we continue?

Optimal stopping theory offers mathematical frameworks for this decision. But pure math assumes you can calculate expected value and compare alternatives numerically. WISER operates where objectives have qualitative dimensions, where "good enough" depends on context algorithms can't capture.

AI systems that iterate, such as agents, optimization pipelines, and automated refinement loops, require explicit stopping criteria because they lack the judgment to know when output is good enough. Without those criteria, they chase marginal improvements that cost more than they're worth. Human practitioners have judgment capacity but may lack frameworks for exercising it well. The gap between "justify resource consumption" and actually making stopping decisions remains unexplored.

We used optimal stopping in deciding that optimal stopping belongs in this appendix. The concept earned enough exploration to preserve, not enough to include in the core method.

Possible future use: Guidance on exercising stopping judgment. Training materials on resource decisions. Integration patterns for AI systems that require explicit stopping criteria.

Individual Requirement Ownership

When questioning requirements, track each one to an individual person with a name. Not to a department or team. Departments perpetuate requirements long after they're obsolete. Individuals can be held accountable for decisions.

Requirements without individual owners are often historical artifacts nobody actually wants anymore. But this is already covered by "People Own Objectives" and the Positions framework. The strategic point is made; this is tactical detail.

Possible future use: Detailed Play guidance for requirement validation exercises.

Distributed Accountability

One perspective says the most critical person is one who can hold the entire problem in their head and make trade-offs. WISER distributes accountability across seven Positions. Are these in conflict?

No. The Sponsor function provides whole-system ownership: understands the full picture, makes integrative decisions. Other functions provide specialized perspectives. Understanding the whole is different from performing every accountability type. The tension resolves through existing structure.

What we're watching: Whether teams get confused about "who owns the whole" in practice. If patterns emerge, we may strengthen the Sponsor description.

Contributing

WISER improves through community learning. If you encounter concepts that might belong here, or develop patterns that address these open questions, the methodology evolves through your contribution.

Share what you find.

PART V

Appendix B: References

Foundational

- **AI First Principles** - Open-source framework governing AI implementation. Available at aifirstprinciples.com. Twelve principles contributed by practitioners, researchers, and builders.
- Wilson, Robb, and Josh Tyson. *Age of Invisible Machines: A Guide to Orchestrating AI Agents and Making Organizations More Self-Driving.* 2nd ed., Wiley, 2025.

Books

AI Safety and Ethics

- Crawford, Kate. *Atlas of AI: Power, Politics, and the Planetary Costs of Artificial Intelligence.* Yale University Press, 2021.
- Russell, Stuart. *Human Compatible: Artificial Intelligence and the Problem of Control.* Viking, 2019.

Methodology and Process

- Beck, Kent. *Extreme Programming Explained: Embrace Change.* Addison-Wesley, 2000.
- Kim, Gene, et al. *The Phoenix Project: A Novel about IT, DevOps, and Helping Your Business Win.* IT Revolution Press, 2013.
- Kim, Gene. *The Unicorn Project: A Novel about Developers, Digital Disruption, and Thriving in the Age of Data.* IT Revolution Press, 2019.
- Poppendieck, Mary and Tom. *Lean Software Development: An Agile Toolkit.* Addison-Wesley, 2003.
- Sutherland, Jeff. *Scrum: The Art of Doing Twice the Work in Half the Time.* Crown Business, 2014.
- Thomas, Dave, and Andrew Hunt. *The Pragmatic Programmer: Your Journey to Mastery.* Addison-Wesley, 2019.

Design Thinking

- Brown, Tim. *Change by Design: How Design Thinking Transforms Organizations and Inspires Innovation.* Harper Business, 2009.
- IDEO. *The Field Guide to Human-Centered Design.* IDEO.org, 2015.
- Norman, Donald. *The Design of Everyday Things.* Basic Books, 2013.

Systems Thinking

- Goldratt, Eliyahu M. *The Goal: A Process of Ongoing Improvement.* North River Press, 1984.
- Meadows, Donella H. *Thinking in Systems: A Primer.* Chelsea Green Publishing, 2008.
- Senge, Peter M. *The Fifth Discipline: The Art and Practice of the Learning Organization.* Doubleday, 1990.

Organizational Learning

- Argyris, Chris. *Organizational Learning: A Theory of Action Perspective.* Addison-Wesley, 1978.
- Dixon, Nancy M. *The Organizational Learning Cycle: How We Can Learn Collectively.* Gower, 1999.
- Gawande, Atul. *The Checklist Manifesto: How to Get Things Right.* Metropolitan Books, 2009.

Risk Management

- Marks, Norman. *World-Class Risk Management.* CreateSpace, 2015.
- NASA. *Risk Management Handbook.* NASA/SP-2011-3422, 2011.

Articles and Papers

- Karpathy, Andrej. "Software 2.0." Medium, 2017. Revisited 2024.
- Meadows, Donella H. "Dancing With Systems." *The Systems Thinker*, 2001.
- Mollick, Ethan. "One Useful Thing" (Substack). Ongoing commentary on AI implementation.

Standards and Guidelines

- IEEE. *Ethically Aligned Design: A Vision for Prioritizing Human Well-being with Autonomous and Intelligent Systems.* IEEE, 2019.
- Partnership on AI. Frameworks for responsible AI development. partnershiponai.org.

Quotes Cited

- Cassie Kozyrkov, quoted in Celonis Face Value interview. "It's never the genie that's dangerous. It's the unskilled wisher."
- Donella Meadows, "Dancing With Systems," 2001. "Before you disturb the system in any way, watch how it behaves."
- Ethan Mollick, Microsoft WorkLab interview, 2023. On treating AI like an intern.
- Jensen Huang, GTC Keynote, 2024. "The new programming language is human language."
- John Gall, *Systemantics*, 1975. "A complex system that works is invariably found to have evolved from a simple system that worked."
- Nassim Nicholas Taleb, *Antifragile*. On subtractive epistemology and removing what is wrong.
- Peter Senge, *The Fifth Discipline*. "The only sustainable competitive advantage is an organization's ability to learn faster than its competition."
- Often attributed to W. Edwards Deming. "Every system is perfectly designed to get the results it gets."

Further Reading

For practitioners seeking deeper context on specific topics:

- **On iteration and experimentation:** Eric Ries, *The Lean Startup*
- **On organizational change:** John Kotter, *Leading Change*
- **On complex adaptive systems:** Dave Snowden, Cynefin Framework
- **On AI implementation patterns:** Chip Huyen, *Designing Machine Learning Systems*
- **On human-AI collaboration:** Ethan Mollick, *Co-Intelligence: Living and Working with AI*

INDEX

Appendix C: Index

A
Abundance Path, 24, 26
Action over Theory, 24, 25, 26
After-Action Review, 106, 193, 194, 195
Agile/Scrum, 224
AI Fails Silently (principle), 28, 32
AI First Principles, 6, 13, 27, 28, 58, 66, 69, 77, 245
AI Inherits Messiness (principle), 28, 32, 34
Ambiguity Is Wisdom (principle), 29, 32
Assumption Inventory, 124, 125, 130, 135, 136
Assumption Testing
Autonomy Graduation Review, 191, 193, 195

B
Build from User Experience (principle), 28, 31, 32
Building vs. Planning, 94

C
Canon Transition Checkpoint, 191, 195
Canons, 14, 15, 31, 32, 35, 37, 86, 89, 95, 164, 174, 188, 193, 207, 212, 241
Champion, 145, 151, 152, 153, 157, 158, 159
Cognitive Layering
Context Comparison Matrix, 72, 145

D
DACI, 99, 168, 208
Data Quality Audit, 127, 130
Deception Destroys Trust (principle), 28, 31, 32, 34
Decompose Incrementally (principle), 29, 32
Design Thinking, 14, 92, 219, 221, 227, 245, 246
Deterministic, 14, 19, 20, 21, 22, 28, 215, 216, 220
Discovery Before Disruption (principle), 29, 31, 34, 240
Document Theater, 29
Drift Alerts, 85
Drift Log, 86
Drift Monitoring, 146, 164, 171, 173, 174, 184
Drift Review, 74, 75, 83, 86, 172, 175

E

Edge Case, 12, 20, 29, 53, 77, 78, 79, 85, 87, 127, 129, 145, 177, 179, 180, 205, 211, 215, 216, 222, ...
EOS, 219, 228, 229
Evolution over Disruption, 25, 26
Expansion Sequence, 153
Experiment Log, 101, 121, 124, 126, 130, 131, 132, 133, 136

F

Friction Map, 43, 44, 45, 46, 47, 89, 107, 110, 113, 114, 115, 117, 120, 208, 236

G

Goodhart's Law, 204
Guardrails, 19

H

Hierarchy of Agency, 76, 77, 86, 105, 144, 164, 167, 184, 202, 240
Human-in-the-Loop, 58, 62, 142, 145, 146

I

Individuals First (principle), 28, 34
Integration Spike, 128
Iterate Towards What Works (principle), 29, 32

J

Justify Resource Consumption (principle), 29, 32, 242

L

Lateral Thinking, 33
Lean/Value Stream, 225

O

Objective Quality Framework, 146
Observation Protocol, 107

P

Paper Prototype, 128
People over Proxies, 25, 26
People Own Objectives (principle), 28, 32, 87, 242
Probabilistic Systems, 13, 14, 20, 22, 220, 225
Process Integrity, 95
Proxy Optimization, 203

R

Rapid Prototyping, 121, 133, 136

Rebuild Decision, 205, 206
Red Team Testing, 177
Reveal the Invisible (principle), 29, 31, 32
Risk Burn-Down, 30, 34, 188, 189, 199, 201

S

Scaffolded Flexibility, 33
Scarcity Path, 26
Six Sigma, 219, 220, 221, 226
Self-Improving Systems, 201, 206
Shadow Mode, 128, 130
Stakeholder Alignment Test
Starter Plays, 7, 33, 34, 89, 207, 208, 209, 212, 234, 236

T

Team Rhythm, 187, 189, 195
Tensions, 34, 87, 89, 91, 92, 96, 97, 98, 129
Trigger Question, 116

U

User Flow Map, 92, 107, 110, 115, 116, 117, 120

V

Value Validation, 137, 141, 146

W

WISER Method, 6, 7, 27, 30, 39, 87
WISER Perspective, 7, 23
Wizard of Oz, 124, 127, 130

GLOSSARY

Appendix D: Glossary

AI First Principles

The twelve constraints that govern AI implementation, numbered in their canonical order.

1. AI Inherits Messiness AI learns from people, so AI systems are inconsistent and operate more effectively with structure. Trying to engineer them to operate like deterministic code results in system failures. The constraint: Define what's prohibited over what's required.

2. AI Fails Silently AI accumulates errors across thousands of interactions before patterns become visible. Traditional systems fail loudly with clear signals; AI fails quietly on repeat. The constraint: Build feedback loops over post-mortems.

3. People Own Objectives AI shouldn't be used in place of human discernment, judgment, or taste. When AI makes mistakes or causes harm, a person should be held accountable, not the algorithm. The constraint: Name the owner.

4. Deception Destroys Trust When AI pretends to be human, people cannot calibrate their expectations, recognize its limitations, and protect themselves from its failures. The constraint: Make AI obvious, not hidden.

5. Individuals First AI industrializes manipulation by personalizing it at scale. Build tools that people control, not tools that control people. The constraint: Prioritize individual agency above efficiency, profit, or convenience.

6. Build from User Experience Without input from end users, AI solutions won't solve real problems. People wrestling with system failures are the ones qualified to design system futures. The constraint: Design systems from lived experience, not distant observation.

7. Discovery Before Disruption Changing systems that aren't understood creates unpredictable failures. Redundancies prevent edge cases and manual steps catch exceptions. Existing inefficiencies are containers of knowledge. The constraint: Identify purpose before simplifying.

8. Ambiguity Is Wisdom Concealing ambiguity removes opportunities for critical judgment. AI produces probabilities that demand judgment, not facts that replace it. The constraint: Surface the probabilities.

9. Reveal the Invisible There's a wealth of ignorance hiding in document theater. Expose what you don't yet understand by learning how to articulate it. The constraint: Pursue what is hard to explain.

10. Iterate Towards What Works Grand plans commit to solutions without validating problems. Iteration tests assumptions and measures impact, revealing what works gradually over time. The constraint: Learn by doing, not planning.

11. Decompose Incrementally Legacy systems carry too much technical debt to replace and are too brittle to automate. AI systems should allow isolated components to decompose naturally. The constraint: Dismantle legacy complexity piece-by-piece.

12. Justify Resource Consumption AI makes it trivially easy to waste resources. What costs pennies to create can cost millions to run. The constraint: Optimize the ratio of value per resource spent.

WISER Canons

The five fundamental practices of WISER. Called "canons" because they represent unchanging fundamentals that are never left behind; when executing the method, you experience them as phases, but you return to earlier Canons as new information emerges. The practices layer and compound throughout the work.

Witness First Canon. Observation reveals what planning conceals. Map the friction people actually feel, forcing solutions to address real problems rather than theoretical ones.

Interrogate Second Canon. Observation finds pain; experiments find causes. Rapid experiments reveal root causes before committing to months of development. Force the system to reveal what's broken.

Solve Third Canon. Experiments find causes; solutions earn trust. Deliver a single, working solution that demonstrates undeniable value. Working software settles arguments.

Expand Fourth Canon. Earned trust enables systematic change toward autonomy. Modularize successful components to solve related problems while maintaining explicit human oversight.

Refine Fifth Canon. Autonomy is not designed; it is grown. AI autonomy increases as reliability is proven. Agency transfers to the system as it proves it can make decisions correctly.

WISER Positions

Named roles that manage the nine inherent tensions in AI implementation. The accountability matters more than the org chart.

Sponsor Position that owns objectives and makes final approval decisions. Manages the tensions of Decision Ownership and Capability vs. Cost. The holistic integrator who understands trade-offs across the entire system.

Architect Position that translates between user needs and technical solutions. Manages the tensions of User Advocacy and System Visibility. Champions user experience when builders optimize for elegance. Owns the Playbook.

Sage Position that brings historical context and institutional knowledge. Manages the tension of Historical Context. Prevents teams from optimizing away critical safeguards.

Scout Position that challenges unvalidated assumptions. Manages the tension of Assumption Testing (short-form: Curiosity). Asks "how do we know that's true?" before consensus overrides evidence.

Smith Position that builds working prototypes and converts plans into software. Manages the tension of Building vs. Planning. Surfaces reality that abstract plans conceal.

Sentinel Position that identifies risks and monitors for failures. Manages the tension of Safety vs. Speed. Watches for silent errors, gaming behavior, and boundary violations. Owns Risk Burn-Down.

Guide Position that maintains process integrity. Manages the tension of Process Integrity. Pushes back when teams want to skip steps under pressure.

General Terms

Abundance Path Strategic orientation that uses AI to tackle bigger problems with more capacity, creating defensible advantage through capability expansion. Contrasts with Scarcity Path.

Action over Theory One of three worldview principles. Trust what can be proven, not what can be planned. Replace planning with proving through small experiments that reveal what works.

After-Action Review Structured learning session after significant events (failures, successes, drift incidents). Produces one clear action; documents learning in Playbook. "Retrospectives with ten actions produce zero change."

Assumption Inventory Catalog of beliefs that haven't been proven. Prioritized list with confidence levels and risk-if-wrong ratings.

Assumption Testing Tension between team alignment and honest verification. Someone must challenge beliefs the team wants to accept. Managed by the Scout.

Autonomy Graduation Review Decision meeting for increasing AI autonomy. Includes graduation proposal, Sentinel risk assessment, Scout challenge, and Sponsor decision with conditions.

Authority See Decision Ownership (Sponsor).

Baseline Metrics captured before any intervention. Required for value validation. Without baseline, you cannot demonstrate improvement.

Boundary Hard limit the AI system must never violate. Includes regulatory requirements, safety constraints, ethical limits, and resource consumption caps. Documented in Playbook; monitored by Sentinel.

Building vs. Planning Tension between planning (which feels productive) and building (which reveals reality). Someone must convert plans into working software. Managed by the Smith.

Canon Transition Checkpoint Formal gate between Canons. Includes validation review, Playbook status confirmation, outstanding items review, and Go/No-Go decision with signatures.

Capability vs. Cost Tension between what we could build and whether it's worth building. Someone must ask "is this worth the cost?" Managed by the Sponsor.

Champion A user within the pilot group who becomes an advocate for the solution. Good champions are respected by peers, have early adopter mindset, and willingly provide feedback. Critical for adoption during Expand.

Cognitive Layering The book's structural principle: each chapter introduces one concept that the next chapter builds on. Readers develop increasing confidence in their understanding rather than accumulating disconnected information.

Context See Historical Context (Sage).

Context Comparison Matrix Side-by-side comparison of pilot context versus expansion context. Shows differences and required adaptations.

Context Fit Assessment Expansion Play that evaluates whether a new context is similar enough to the pilot context for direct deployment. Surfaces differences that require adaptation before rollout. Part of the Expand Plays.

Curiosity See Assumption Testing (Scout).

DACI Framework Decision-making model: who Drives, Approves, Contributes, and is Informed for each decision type. Prevents ambiguity about who owns what within Positions.

Data Quality Audit Experiment type that samples real data to check completeness, accuracy, format consistency, and edge case frequency. Used before assuming training data or inputs are usable. Duration: 2-5 days.

Decision Ownership Tension between collaborative input and accountable output. Someone must own the decision when the team can't agree. Managed by the Sponsor.

Deterministic Systems Systems that produce predictable, repeatable outputs given the same inputs. Traditional software is deterministic. AI systems are probabilistic.

Deployment Gating Expansion Play that provides go/no-go criteria before deploying to a new context. Ensures baseline metrics, success criteria, rollback capability, and user preparation are in place. Part of the Expand Plays.

Document Theater Documentation that describes ideal processes rather than actual practice. Term from Dave Thomas, co-author of *The Pragmatic Programmer*. "A wealth of ignorance hiding in document theater." Witness exposes document theater by observing real work.

Drift Gradual degradation in AI system performance or behavior over time. Can result from changing data patterns, model decay, or shifting user behavior. Detected through monitoring; addressed through Refine practices.

Edge Case Scenario that occurs infrequently but reveals system limitations. "Edge cases that were rare become common" at scale. Handling edge cases distinguishes pilots from production systems.

Escalation When AI stops and requests human input because a situation exceeds its defined boundaries or confidence thresholds. Clear escalation triggers are defined in the Hierarchy of Agency.

Evolution over Disruption One of three worldview principles. Rebuild the system while it runs, not shutting it down for a rewrite. Advance what's broken without stopping what works.

Empathy See User Advocacy (Architect).

Expansion Sequence The prioritized order for rolling out a validated solution to new contexts. Based on similarity to pilot, risk level, champion availability, and learning value.

Expansion Readiness Check Expansion Play that determines whether your current deployment is stable enough to replicate. Answers the question: should we expand now, or stabilize first? Part of the Expand Plays.

Execution See Building vs. Planning (Smith).

Experiment Log Running documentation of experiments and results. Captures hypothesis, test design, success/failure criteria, results, and next steps.

Friction Map Visual documentation of where work breaks down. Shows friction points with frequency, time impact, and severity. Created during Witness.

Gaming When users or systems exploit metrics in ways that satisfy measurements while undermining actual objectives. A key risk that Sentinel monitors.

Goodhart's Law "When a measure becomes a target, it ceases to be a good measure." The moment you tell an AI system to optimize a metric, you create incentive to game that metric. For every metric you optimize, identify how it could be gamed, then measure the game.

Graduated Autonomy The principle that AI autonomy increases as reliability is proven. Three stages: Tier 3/supervised (human reviews all actions), Tier 2/semi-autonomous (spot-checks), Tier 1/fully autonomous (auto-approve with monitoring). Graduation moves from Tier 3 toward Tier 1 as reliability is proven. Autonomy is earned through demonstrated reliability, not designed initially.

Graduation The process of increasing AI autonomy based on demonstrated reliability. Systems graduate from supervised to semi-autonomous to fully autonomous as they meet defined thresholds documented in the Playbook.

Guardrails Constraints that prevent AI from taking prohibited actions. Tested through Red Team exercises. Guardrails can be exploited, gamed, or gradually eroded through drift.

Hierarchy of Agency A spectrum defining what AI can decide independently, what requires human approval, and escalation triggers. Documented in the Playbook and evolves as the system demonstrates reliability.

Historical Context Tension between moving fast and understanding what you're changing. Someone must know why things work the way they do. Managed by the Sage.

Human-in-the-Loop Operational protocol where AI produces primary outputs and humans review before execution. The AI decision is what gets acted upon (after human approval) or corrected (if not). Used during pilots and early deployment to catch errors, build training data, and develop trust. Oversight reduces as system graduates to higher autonomy. Contrast with Shadow Mode, where AI observes but doesn't drive execution.

Integration Spike Experiment type that builds the smallest possible working connection to external systems. Reveals undocumented API behaviors. Duration: 3-5 days.

Integrity See Process Integrity (Guide).

Iteration Returning to earlier Canons when new information emerges. Expected practice in WISER, not process failure. Discovery in Expand may trigger return to Interrogate; complexity in Solve may require return to Witness.

Lateral Thinking Solving problems by approaching them from unexpected angles rather than following established paths. Teams with lateral thinkers tackle hard problems with creative freedom, creating competitive moats that efficiency-focused competitors cannot replicate.

Objective Quality Framework Tests to ensure objectives are actionable and measurable. Validates clarity, measurement capability, trade-offs, gaming resistance, root cause understanding, autonomy enablement, and scope.

Observation Protocol Structured approach for Witness sessions. Includes session structure (before, during, after), question bank by role, and documentation standards.

Paper Prototype Experiment type that sketches a workflow and walks users through it manually. Tests whether users understand the flow before building interfaces. Duration: 1-3 days.

People over Proxies One of three worldview principles. Value the experts doing the work over the ones documenting it. Treat the humans doing the work as the primary source of truth.

Pilot A controlled deployment to a limited user group to validate that a solution works before broader rollout. Includes baseline metrics, success criteria, rollback plan, and user preparation.

Pilot Purgatory When successful pilot projects fail to scale to production. Common failure mode where organizations repeatedly prove value in pilots but never achieve operational deployment.

Playbook Living documentation that captures objectives, constraints, risks, decisions, and learning. The organizational memory that survives chaos. Prevents the insights that drove success from fading before they can be scaled. Not a document you write once, but a system that evolves with every cycle.

Plays Tactical guidance that transforms the abstract WISER framework into execution for specific contexts. Modular components (mapping techniques, frameworks, templates, cadences) that can be used as-is, adapted, or replaced with alternatives.

Positions Named roles that manage the nine inherent tensions in AI implementation. The Starter Plays define seven Positions (Sponsor, Architect, Sage, Scout, Smith, Sentinel, Guide) that can be combined or distributed based on team size. The accountability matters more than the org chart.

Probabilistic Systems Systems that produce probabilities rather than certainties. AI systems behave probabilistically: they fail silently, learn from mess, and drift without warning. Traditional project management designed for deterministic systems struggles with probabilistic behavior.

Process Integrity Tension between rigor and urgency. Someone must hold the line on methodology when pressure mounts. Managed by the Guide.

Proxy Optimization When AI improves the metric you gave it while the actual objective declines. Example: You measure response time; AI produces fast but incomplete answers. The solution is measuring both proxy and objective simultaneously.

Rapid Prototyping Building quick, disposable artifacts to test assumptions. "Ship ugly, validate fast, throw away without guilt." Can be paper prototypes, wireframes, or working code. The goal is learning, not polish.

Rebuild Decision Framework for evaluating whether to iterate or rebuild from scratch. Default is iterate. Before considering a rebuild, prove you understand the system by manually replicating the process.

Red Team Testing Adversarial testing to find where AI guardrails fail. Tests for boundary violations, boundary exploitation, metric gaming, and gradual drift. Required before increasing autonomy.

Risk Burn-Down Continuous process of identifying highest-priority risks (scored by probability, impact, and detection difficulty), systematically reducing them, and moving to the next. Synthesizes Theory of Constraints (address highest priority first) with continuous assessment.

Rhythm Plays Meeting Plays that keep teams aligned, risks visible, and blockers resolved. Includes Team Rhythm (weekly alignment), Canon Transition Checkpoint (formal gates between Canons), and After-Action Review (structured learning after significant events). The operational heartbeat that prevents drift.

Rollback The ability to revert to previous state if a deployment fails. Required capability before any expansion or autonomy graduation. "Can you revert within hours if something breaks?"

Rollout Systematic deployment to additional contexts after pilot success. Sequenced based on similarity to pilot, risk level, champion availability, and learning value.

Safety See Safety vs. Speed (Sentinel).

Safety vs. Speed Tension between pressure to ship and unintended consequences. Someone must watch for silent failures. Managed by the Sentinel.

Scaffolded Flexibility WISER's balance between structure (prevents chaos) and flexibility (enables discovery). The Canon sequence provides scaffolding; iteration back to earlier Canons provides flexibility.

Scarcity Path Strategic orientation that uses AI to do the same work with fewer people. Produces short-term efficiency but long-term stagnation. Contrasts with Abundance Path.

Self-Improving Systems Framework where AI optimizes how it works, but humans own what it's working toward. AI can adjust confidence thresholds, processing efficiency, and routing logic. AI cannot adjust objectives, hierarchy of agency levels, boundary constraints, or escalation triggers.

Shadow Mode Experiment type where AI runs in parallel with the existing human process. AI outputs are logged but not executed; human decisions remain primary. The comparison measures AI accuracy against human baseline to validate readiness before giving AI real responsibility. Duration: 1-4 weeks. Contrast with Human-in-the-Loop, where AI outputs are executed after human approval.

Stakeholder Alignment Test Experiment type that validates whether key stakeholders actually agree on objectives, constraints, and success criteria. Surfaces political risk early. Duration: 1-2 days.

Starter Plays The collection of Plays provided in Part IV. Proven starting points for teams beginning their WISER practice. Designed to be adapted, renamed, combined, or replaced as practitioners gain experience.

System Visibility Tension between system complexity and team comprehension. Someone must make invisible complexity visible. Managed by the Architect.

Stewardship See Capability vs. Cost (Sponsor).

Team Rhythm Weekly meeting that keeps the team aligned, risks visible, and blockers resolved. The one rhythm you cannot skip. Includes Playbook check-in, progress review, risk burn-down, blocker resolution, and priority setting.

Trigger Question Before mapping any user flow: "What triggered the user into this journey?" If you cannot answer, you should not be mapping. No trigger means no user means no solution worth building.

Translation See System Visibility (Architect).

User Advocacy Tension between technical elegance and user reality. Someone must champion users when builders optimize for what's technically interesting. Managed by the Architect.

User Flow Map Process flow showing how users move through the system. Swimlane diagram showing user journey with handoffs.

Value Validation Process of demonstrating that a solution delivers measurable improvement over baseline. Includes before/after metrics, qualitative feedback, and confidence level assessment.

Vibe Coding Writing code by intuition rather than detailed specification, often with AI assistance. The practitioner iterates based on what feels right, letting the code emerge through rapid feedback loops. Useful during Interrogate and early Solve when exploring solution shapes. Not the same as Rapid Prototyping: you can vibe code a production feature or carefully plan a paper prototype.

Wizard of Oz Experiment type where a human performs the "AI" function behind the scenes without users knowing. Tests whether users want the capability before building it. Duration: 1-2 weeks.

Workaround Unofficial process users create to bypass system limitations. Workarounds are signals, not problems. They reveal unstated requirements nobody thought to document.

GLOSSARY

www.ingramcontent.com/pod-product-compliance
Lightning Source LLC
LaVergne TN
LVHW072124060526
838201LV00069B/4964